NEW DEVELOPMENTS IN CHEMISTRY AND BIOCHEMISTRY

New Developments in Chemistry and Biochemistry

G. E. Zaikov and V. M. M. Lobo
Editors

Nova Science Publishers, Inc.
New York

Senior Editors: Susan Boriotti and Donna Dennis
Coordinating Editor: Tatiana Shohov
Office Manager: Annette Hellinger
Graphics: Wanda Serrano
Editorial Production: Marius Andronie, Vladimir Klestov, Matthew Kozlowski, Tom Moceri, Anthony T. Sovak and Maya Columbus
Circulation: Ave Maria Gonzalez, Vera Popovich, Luis Aviles, Melissa Diaz, and Jeannie Pappas
Communications and Acquisitions: Serge P. Shohov
Marketing: Cathy DeGregory

Library of Congress Cataloging-in-Publication Data
Available Upon Request

ISBN: 1-59033-653-4.

400 Oser Ave, Suite 1600
Hauppauge, New York 11788-3619
Tele. 631-231-7269 Fax 631-231-8175
e-mail: Novascience@earthlink.net
Web Site: http://www.novapublishers.com

Printed in the United States of America

CONTENTS

Chapter 1

PHYSICOCHEMICAL CRITERIA FOR ESTIMATING THE EFFICIENCY OF BURN DRESSINGS

Klara Z. Gumargalieva and Gennady E. Zaikov*
N.N. Semenov Institute of Chemical Physics, RAS
* N.M. Emanuel Institute of Biochemical Physics, RAS
4, Kosygin str., Moscow 119991, Russia

ABSTRACT

Based on theoretical and experimental data, it was found that maximal sorptional ability of a burn dressing equals to the free volume of the dressing material calculated from the value of the material density. Kinetic parameters were determined from the sorption curves. These parameters help in predicting behavior of burn dressings. The criteria of estimating efficiency of first aid burn dressings were formulated.

Keywords: sorptional ability, free volume, model liquid, blood plasma medium, kinetic parameters, penetrability, porous space, adhesive strength, criteria of the efficiency, first aid

1 INTRODUCTION

The principal medical treatment of burns is the use of dressings, which often worsen the effects of the injury.

It is difficult to estimate the effectiveness of the new burn dressings. Their physicochemical properties are not usually presented in literature.

This article firstly discusses this subject. The authors address the complexity of physicochemical methods of analysis in order to create criteria for efficient dressings for a burn wound surface.

2 MODERN SURGICAL BURN DRESSINGS

Dressings for wounds and burns must primarily be protective, sorptional and atraumatic. In currently used dressings these properties are provided by a multilayered structure or structural modifications.

Different classifications of the dressings can be found in literature: by material, by construction or by functions [1 - 3].

The dressings applied in the modern treatment of wounds and burns are subdivided into three groups according to material of the layer, sorbing the exudate of wounds; i.e., material of animal origin; synthetic, foamy polyurethane; material of vegetable origin.

2.1 Dressings Based on Materials of Animal Origin

Typical dressings for this group are collagen sponges, which are porous. Beside hydrophilic properties, collagen sponger provide higher sorption of liquid (in the range of 40 - 90 g/g) [1, 4 - 9].

The patent literature describes in detail the methods of obtaining collagen dressings for wounds and burns. In the form of sponges and felt [10 - 13] based on materials of animal origin also include these made from biological artificial leathers based on liophilized body's and swine cutis, produced as plates 0.5 - 0.7 mm thick. However these materials possess lower sorptional capacity than collagen dressings.

Dressings called "cultivated cutis" are also obtained from the epithelia of cells of the patient himself [13].

The shortcoming of biological artificial leathers or biodressings their expense and their inability to retain, as a rule, their properties on storage.

2.2 Dressings Based on Synthetic Materials

The demands inexpensive raw materials for the production of dressings for wounds and burns led to the production based on synthetic polymeric materials, particularly cellular polyurethane [14 - 19].

Cellular urethanes, intended for medical purposes, are synthesized on the basis of toluilenediisocyanate and polyoxypropyleneglycole [20].

The dressings based on polyurethanes have a pores distribution of about 200 - 300 pores/cm^2, and allow the regulation of the number and size of pores in layers [21]. On the whole, dressings from this group are prepared double layered, the outside layer being of higher density in order to prevent liquid evaporation and penetration of microorganisms. In rare case, these dressings are made homogeneous by all the thickness.

The influence of the pores sizes on sorptional properties of polyurethane sponges is described in the paper [22]: macroporous sponge with the pore size from 200 up to 2000 μm is completely nourished by exudate under pressure only. In this case through pores should possess the sizes of several micrometers.

Table 1. Characteristics of Dressings in Brief

No	Name	Company	State	Structure	Composition
Group 1					
1.	Collagen burn dressing	"Helitrex"	USA	Dressings uniform by thickness, dense, pores of 0.01 mm size. It has gauze cover	Collagen
2.	Collagen sponge	"Helitrex"	USA	Similar to No. 1, differs by big radius of pores, formed by fibril weaving of cylindrical form preferably	Collagen
3.	Collagen dressing	"Bayer"	Germany	Friable dressing, possesses rough porous structure with pores-holes sizes from 1.5 to 0.1 mm	Collagen
4.	Burn curative dressing	"Combutec-K 11"	USA	Dressing of large-porous structure with pores size from 1 to 0.05 mm. Pores are of cylindrical form preferably formed by fibrils weaving of collagen	Collagen
5.	Biological dressing	"Corretium-2"	USA	Dense, pressed plate. Fibrillar structure is observed in dense layers	Collagen
6.	Biological dressing	"Corretium-3"	USA	The same	Collagen
Group 2					
7.	Compositional burn dressing	"Biobrant"	USA	Double-layered elastic, porous dressing, consists of the upper layer of 0.01-0.005 mm and flexible fabric nylon base. It represents combination of hydrophilic components with elastic silicon films	Silicon, the main layer - from polyamide
8.	Synthetic dressing	"Epigard"	USA	Double-layered elastic porous dressing. Upper layer is dense, non-porous 0.2 mm thick	The main layer - from polyurethane, the upper one - polypropylene
9.	Synthetic burn dressing	"Syncrite"	ChSSR	Single-layered dressing on gauze base with through large pores, medium flexible	Polyurethane

10.	Synthetic wound dressing	"Syspurderm"	GDR	Dressings of homogeneous composition, with different pores distribution: upper layer is 0.1 mm thick, possesses small porous structure with pores of 0.01 mm size; lower layer, adjoining wound possesses large pores of 0.05 mm size. The dressing is "elastic", accepts a form badly	Polyurethane
11.	Synthetic wound dressing	"Farmexplant"	PBR	Antiseptic double-layered dressing. Main polyurethane layer possesses pores of 0.1-1.5 mm. Upper layer is 0.1 mm thick, more dense, non-porous	Polyurethane
12.	Atraumatic caproic dressing		USSR	Large-cellular dressing on basis of weaved nylon	Polyamide
Group 3					
13.	Cover for wound, burns	VNIImedpoly-mer "Algipor"	USSR	Wound large-cellular dressing, homogeneous by its composition	Alginic acid salts
14.	Needle-pierced fabric		USSR	Porous cotton balling dressing with atraumatic layer	Cellulose
15.	Wound non-adhering dressing	"Bayersdorf"	FRG	Three-layered dressing of plaster type with tricot lower layer. Dressing is of the sandwich type: upper layer-creped paper, main part is cotton balling, lower layer-tricot network. Atraumatic action is provided by the effect of dressing "bending" (tunneling effect)	Main and upper layers - from cellulose, lower one is a film of Dacron or nylon
16.	Wound absorbing dressing	"Johnson-Johnson"	USA	Three-layered dressing with perforated lower and upper layers 0.01 mm thick, main part is cotton balling, porous	Main layer - from cellulose, external layers - from polypropylene
17.	Haemostatic		Sweden	Double-layerd dressing with perforated lower layer, sewed to the main layer	Viscose main layer, atraumatic one - from polyethylene
18.	Wound dressing	"Mesorb"	France	Cotton balling or viscose dressing, creped paper - lower and upper layers	Cellulose

19.	Surgical dressing	"Kendall"	USA	Similar to No. 16 with cellulose base and atraumatic synthetic lower layer	Cellulose
20.	Dressing with perforated metallized layer		Germany	First aid dressing with hydrophobic layer and lower metallized layer. Internal layers represent non-fabric pressed layer of creped paper	Min layer - from cellulose, lower layer is aluminum spray-coated
21.	Dressing. lower layer is not metallized		Germany	Similar to No. 20	Cellulose with spray-coated lower layer
22.	Non-adhesing dressing	"Switin"	ChSSR	Dense cotton balling dressing, lower and upper layers are non-fixed nylon networks	Cellulose and polyamide
23.	Series of experimental dressings with various quantitative viscose-cotton composition		USSR	Cellulose or viscose dressings with atraumatic layer	

Other polymeric materials (polyvinyl chloride, nylon, etc.) are applied as the sorbing layer apart from polyurethanes [23 - 26]. This group should be added by a compositional burn dressing, based on a silicon film, polyamide network and hydrophilic admixture, produced by Hall Woodroof Co. (USA) [13]. Polyurethane covers with atraumatic lower layer from polyglycolic acid may be considered as variety of compositional dressings [27]. It is characteristic of dressings from this group that they preserve their high strength properties at the absorption of the wounds exudate.

A two-component protective dressing "Hydron" was recently applied in the treatment of burns. It is a film, formed on the wound, a powder of poly-2-hydro-oxyethylmetacrylate dissolves in polyethyleneglycol 400 [2, 28]. Although they possess good protective properties, "Hydron" dressings have low strength and sorptional capacity.

2.3 Dressings Based on Materials of Vegetable Origin

A large number of burn dressings are so-called "cotton-batting" ones, based on cellulose, viscose or a combination of the two [29 - 33]. These dressings differ from each other by structure and composition of the upper and lower layers. Most often the sorptional layer from cellulose is used in complex dressings. Such dressings are usually layered, with the separate layers being produced from either the same material or from different ones, and may be a fixed mechanically or by using thermoplastic material. To decrease their adhesion to the wound surface, the lower layer is produced from various fabric and non-fabric materials (perforated Dacron, polypropylene, pressed paper, metallized fabric material, etc.). The total sorptional ability of these dressings is defined by hydrophilicity and porosity of the basic material and is usually equal 15 - 25 g/g.

Papers [1, 34, 35] show the dressing action data on wounds and burns of dressings based on another vegetable material-derivatives of alginic acid. Typical "Algipor" specimens are used based on mixed sodium-calcium salt of alginic acid as spongy plates of about 10 mm thick with high absorption ability (Table 1).

3 Selection of Properties of Burn Dressing Being Tested

The literature data showed that burn dressings, particularly the first aid ones, must perform three main functions [1, 2, 36, 37]:

1. Absorb the wound exudate, which contains metabolism products and toxins.
2. Provide optimum water, air and heat exchange between the wound and the atmosphere.
3. Protect the wound from the penetration of microorganisms from the air.

Moreover, the burn dressing must be removable from the wound without further injury to the patient.

Therefore the following properties of burn dressings are studied to determine their efficiency.

3.1 Sorption-Diffusion Properties

Sorption-diffusion properties of dressings are extremely important, because they determine the performance of the three main functions of dressings.

Water Absorption

Water is the main component of exudate of wounds. At present, there is no opinion on how fast and in what degree the dressing must absorb the exudate enough to clean the wound from toxins and metabolism products while, the same time, keeping wound wet, enough to present the removal of water from healthy tissue [1, 2, 36, 37].

Air Penetrability

Sufficient air must be allowed to penetrate the dressing since increased oxygen concentration helps the healing process.

Vapor Penetrability

Vapor penetrability of skin of a healthy man is 0.5 $mg\times cm^{-2}hour^{-1}$ [38]. In absence of technical data it may be concluded that high vapor penetrability will lead to "drying" of dressing with a corresponding change of surface energy on the dressing-wound interface. On one hand this will promote undesirable removal of water from the tissues. On the other hand, it may cause the dressing to come off the wound. Low vapor penetrability of dressing will lead to the accumulation of liquid under the dressing that may cause edema.

Penetrability with Reference to Microorganisms

Penetration of microorganisms through the dressing must be blocked to prevent infection.

3.2 Adhesive Properties

The adhesive properties of dressings determine their ability to stay attached to the wound. Thus, the surface energy of the dressings surface facing the wound must always be lower than that of the wound surface.

3.3 Mechanical Properties

The following two mechanical properties are important for dressings: flexural rigidity and strength at break.

The former defines the ability of the dressing to accept wound profile; the latter is important since it allows the dressing to be removed from the wound completely without breaking.

4 Methods of Investigation of PhyslcoChemical Properties of Burn Dressings

4.1 Determination of the Materials Porosity

The porosity of materials (the relation of pore space volume to total volume) is determined according to two methods:

1. Measuring density:

$$Q = 1 - \frac{\rho}{\rho_0}, \tag{1}$$

where **Q** is the material porosity; ρ is the observable density; ρ_0 is density of the material forming porous medium.

The value of ρ is determined by weighting of a sample with the known geometrical sizes. The value of ρ_0 is determined similarly for the samples, pressed by 500 GPa.

2. From photos obtained by microscope we get

$$Q = \left[\frac{S_{pores}}{S_0} \right]^{3/2}, \qquad (2)$$

where S_{pores} and S_0 are total surface of pores and general surface of the material in the field of vision of the microscope, respectively.

4.2 Determination of Sizes and the Amount of Pores

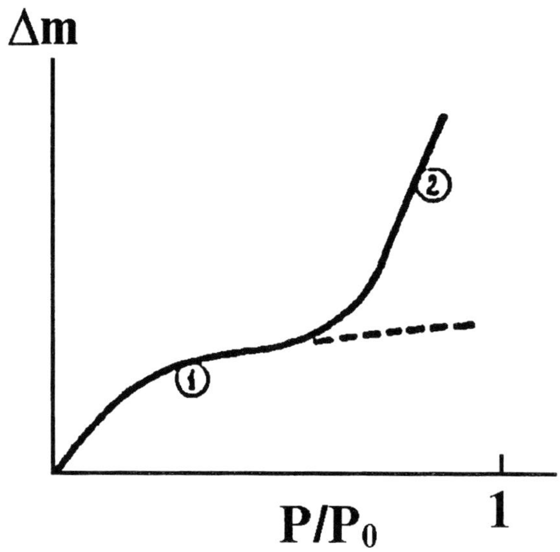

Figure 1. Sorption isotherm of low-molecular liquids by microporous materials: 1 - real dissolving of liquid by the material; 2 - condensation of liquid in the material microporous; m - sorbed liquid mass. P/P_0 is relative pressure of liquid in thermostating vessel (P_0 is pressure of saturated vapor of liquid in the present conditions).

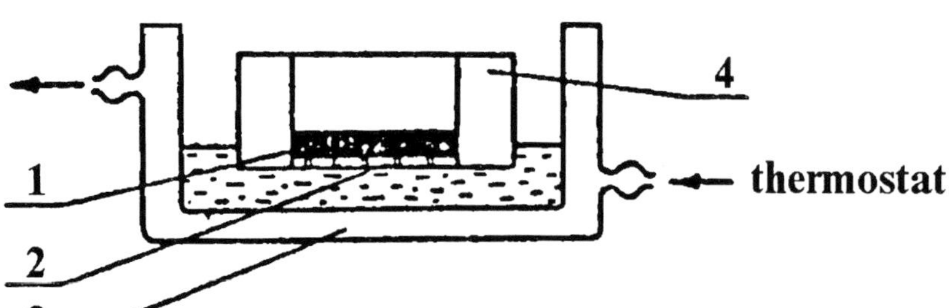

Figure 2. The scheme of device for determination of absorbing ability of the porous materials: 1 - sample; 2 - perforated plate; 3 - thermostated camera: 4 - float.

The amount and sizes of pores are determined with the help of MIN-10 microscope in reflected light. The curve of pores distribution by radii is calculated.

4.3 Estimation of Surface Energy on the Material-medium Interface

Surface energy on the material-medium interface is estimated using the wetting angle of the material surface by the medium. A drop of liquid is applied to the surface of the material, and the angle is measured between the tangent at the base of the drop and the material surface. The wetting angle is determined using horizontal microscope HM. The accuracy of angle measurement does not exceed ± 1°.

4.4 Determination of Sorptional Ability of Materials

The total amount of liquid sorbed by tile material, includes the liquid in macropores with the size over 0.1 μm, in micropores with the size smaller than 0.1 μm and in the material matrix (dissolved liquid).

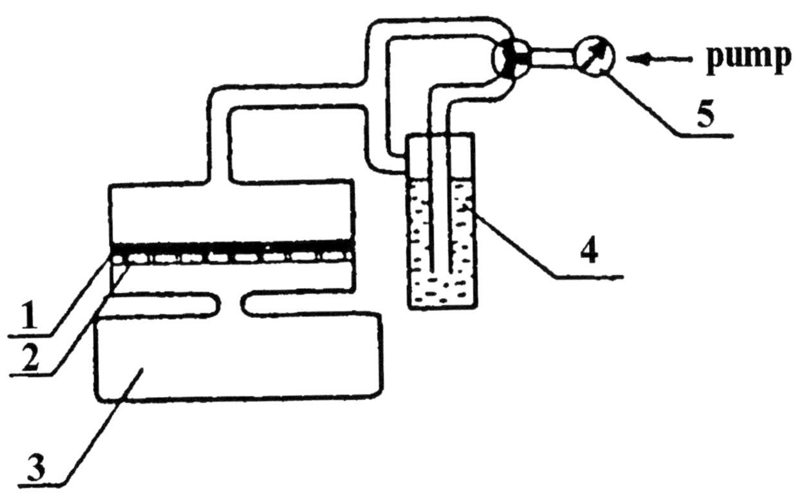

Figure 3. Device for determination of air penetrability of materials: 1 - sample: 2 - perforated plate; 3 - polyethylene sack; 4 - manometer; 5 - pressure controller.

The amount of dissolved liquid, and of liquid filling the micropores, is calculated from the dependence of the amount of sorted liquid on pressure of vapors of that liquid over the sample (isotherms of sorption).

The sorption isoterms for the materials with micropores possess S-type forms (Fig. 1). The first part of the curve is connected with the real desolving of liquid, the second one-with the condensation of liquid in micropores.

The maximal sorption (the amount of liquid, really solved and filling micro- and macropores) is determined with the help of the device, shown in Fig. 2. The device represents a vessel with liquid medium, in which a float of special perforated square construction is placed. The float construction is calculated to prevent its sinking. This requires that the liquid medium docs not penetrate through the perforation of the square, but forming meniscus on the side of the square facing the porous interlayer. The change of the mass of porous material is determined from the immersion of the float with the sample. It is performed with tile help of a horizontal microscope HM.

4.5 Determination of Air Penetrability of Burn Dressings

Air penetrability (the volume of the air, passed through the specific surface during the specific time) was determined with the help of device specially designed for this purpose. The device is a cylindrical cell with perforated plate supporting the sample (Fig. 3). The air was passed through the cell with the help of air compressor, equipped with manometer and pressure controller. We measured the time required to fill with air a polyethylene 45 1 volume sack.

The sample of a round form was cut with the help of mould. Then the sample was placed on the perforated plate of the cell. The sample was pressed by screwing the air was compressed through the cell. The time of polyethylene sack filling was measured with the help of stop-watch. The method allows to determine the air penetrability of dry or wet materials.

4.6 Determination of Adhesion of Burn Dressings

Adhesion of burn dressings was investigated on the device, representing the modification of the device, described in the paper [39].

A 1 mm thick fiber glass plate, covered with three layers of medical gauze, was placed into the fiber glass cell having a working surface of 3 × 10 mm. The cell was filled of 5 ml of the whole blood added by 1 ml of 2% thrombin. The dressing tested was placed on the plate surface 1 min after. The cell with the sample was placed into thermostat at 37°C for 24 hours. The sample removal was performed at 90° angle with respect to the surface of the tested material.

4.7 Determination of Vapor Penetrability of Burn Dressings

Vapor penetrability (the mass of water, passed through the specific surface during the specific time) was determined with the help of the device described in [40]. Glass vessel was filled of a definite amount of liquid, for example water or water solution of sulfuric acid. This amount provided the definite relative humidity. The investigated sample was placed on the vessel surface, metal ring was set and pressed to the vessel by special clamp. The vessel with the contents was weighted and placed into desiccator with dryer at 37°C. After definite period of time the vessel was taken out from the desiccator, weighted and then put back to the desiccator. The amount of water, passed through the sample, was determined by the mass losses of the vessel contents. The vessel of 40 mm diameter and 20 mm high was used in the experiments.

Table 2. Composition of Edema Liquid and Blood Plasma, g/ml.

Components	Edema liquid	Blood plasma
Urea	5.1×10^{-4}	5.5×10^{-4}
Sugar	5.8×10^{-6}	11.0×10^{-6}
Protein	3.4×10^{-2}	7.2×10^{-2}
Salts	1.0×10^{-2}	1.0×10^{-2}
Albumin/Globulin	3.9	1.5

5 Experimental Results and Discussion

5.1 Determination of Sorption Ability of Burn Dressings

At applying the dressings on the burn wound there occurs first the wetting of the surface layer of the material and then sorption of the wound exudate into the dressing volume. In this connection it is necessary to answer the following two questions:

1) What are the components of the exudate of wound and burns, being able to sorb by the material, and what is the way of sorption?
2) What is the maximum sorption of the separate components of the exudate by the dressing material?

The first question had not yet been addressed in the published literature. In regard to the second question, the maximum water sorption of different materials as follows was previously determined [41]. The sample was immersed into the water, dried fast by filter paper and then weighted. Such method did not allow one to measure the sorption kinetics, and its accuracy of the maximum sorption was low.

That is the reason why we have worked out the device for continuous measurement of sorption, which solved the above mentioned shortcomings.

The exudate of the wounds contains water, salts, proteins, cells the damaged cells and various low- and high-molecular substances in relatively lower amounts. Table 2 shows the approximate composition of edema liquid in the burn wound. The composition of edema liquid changes in dependence on the burn degree: the worse the bum is, the higher the content of protein and the lower albumin to globulin ratio [3]. Similar data for the blood plasma are shown for comparison in Table 2.

Sorption of wounds exudate may proceed via filling of micro- and macropores, or dissolving in the material matrix.

Let consider sorption of different components of the wound exudate by the dressing material.

Water fills pores and dissolves in the material matrix. Water solubility is defined by the material hydrophility.

The solubility of water, salts and other low molecular substances in polymers is the subject to the following rules:

- — solubility in hydrophilic polymers is defined by size and charge of low molecular substance;
- — in hydrophobic polymers it is defined by vapor elasticity (the higher vapor elasticity is, the higher is the solubility) [42].

Protein fills the pores of the size up to 10^{-2} m and may dissolve only in hydrogels of "Hydron" type with water content over 30% by mass.

Cells fill only open pores of the size over 0.1 - 0.2 μm.

Solubility of Water in Polymers

As above mentioned modern burn dressings represent heterogeneous materials, usually consisting of several layers. The upper one having the air is, as a rule, more hydrophobic and less porous than the others.

Solubility of water in this layer will define its evaporation from the dressing surface and the heat exchange between the wound and the surrounding. The information about solubility of water in various polymers (Table 3) are shown in the paper [43].

Table 3. Solubility of Water in Various Polymers.

Polymer	Solubility, 10^2 g/g	T, K
Cellophane	40	303
Viscose fiber	46	303
Cotton	23	303
Cellulose diacetate	18	303
Cellulose triacetate	11.5	303
Polycaproamide	8.5	303
Polyethyleneterphthalate	0.3	303
Polydimethylsiloxane	0.07	308
Poly(2-oxyethylmethacrylate)	40*	310
Polypropylene	0.007	298
Polytetrafluoroethylene	0.01	293
Polyethylene (= 0.923)	0.006	298
Polyurethane	1*	298
Polyvinylchloride	1.5	307

Note: * is measured by the authors.

Solubility of water was determined by the sorption method. Extreme values of sorption at the definite pressures of water vapors were calculated from the sorption curves, and then the sorption isotherms were constructed using the method described in the paper [44].

Extreme values of solution $\varphi^{\infty}_{H_2O}$ at the saturation pressure were determined by extrapolation of φ_{H_2O} to $P/P_s = 1$. The value $\varphi^{\infty}_{H_2O}$ equals to the solubility of water in polymer.

Maximum Sorption Ability of Burn Dressings

Modern burn dressings represent large-porous or fibrillar heterogeneous materials, possessing high free volume. At the contact with the wound the exudate will fill the free volume of the dressing, filling degree being defined by hydrophility of the material, the size and geometry of the free volume fraction.

Theoretical

Consider the filling process of the entire free volume of the material by liquid medium. The maximum sorption of the medium by the dressing will be calculated in the following way. Mass of the medium sorbed by the dressing material, equals to:

$$\mathbf{m_c} = \mathbf{V_\infty \rho_c} - \mathbf{V_0 \rho_0}, \tag{3}$$

where $\mathbf{V_\infty}$ and $\mathbf{V_0}$ are the volumes of the dressing after and before sorption of the liquid medium, respectively; ρ_c and ρ_0 are the density of the liquid medium and the dressing material, respectively.

Maximum sorption of the liquid medium in this case is

$$\mathbf{C_c^\infty} = \frac{\mathbf{m_c^\infty}}{\mathbf{m_0}} = \frac{\mathbf{V_\infty \rho_c}}{\mathbf{V_0 \rho_0}} - 1, \tag{4}$$

where $\mathbf{m_0}$ is the initial material mass.

Two particular cases are possible.

1. There is no swelling of the polymer during sorption, i.e. liquid medium fills the free volume space only. In this case $\mathbf{V_\infty} = \mathbf{V_0}$ and

$$\mathbf{C_c^\infty} = \frac{\rho_c}{\rho_0}, \tag{5}$$

2. For materials possessing low density, $\rho_0 < 0.1$ g/cm^3

$$\mathbf{C_c^\infty} = \frac{\rho_c}{\rho_0}, \tag{6}$$

Maximal Sorption of Water by Burn Dressings

Sorption of water by burn dressings is measured using a device developed for this purpose by the authors.

Experiments were performed in the following way. First, the weights of different mass were placed on the perforated plate of the device and the relative device immersing to the water was measured with the help of a horizontal microscope HM. Calibrating curve was represented in "weights masses - depth of device immersing" coordinates in the units of eyepiece of microscope. Angle coefficient equals 0.70 ± 0.02 g/unit.

Subsequently, the sample of a dressing was placed into the device, and the depth of the device immersing during time "**h**" was measured. The mass of the medium sorted by the material was calculated from the correlation:

$$\mathbf{m_c} - 0.70\mathbf{h}. \qquad (7)$$

The extreme value of the sorted medium mass was determined at **t** → ∞. Maximal sorption of the medium by the material was calculated from equations (5) and (6).

Table 4 shows experimental and theoretical (calculated from (7)) values of $\mathbf{C^{\infty}_{H_2O}}$ and values of ρ_0, determined experimentally and used for theoretical calculation.

Good correlation was observed between the experimental and theoretical values of $\mathbf{C^{\infty}_{H_2O}}$ for the majority of dressings. This shows that practically the entire free volume is filled by liquid medium at the contact of dressings with water.

The exception is the "Algipore" dressing, the large pores of which become denser on filling of water entering because of collapse of the pore walls. In the end this leads to the decrease of the total volume of the dressing. Liquid medium may fill not the whole volume of dressings, if the material is sufficiently hydrophobic and poorly wetted with water.

To test this assumption, seven collagen materials were investigated, which differed in the production method.

We investigated: density, maximal water sorption, wetting angle and heat effect of water sorption by the material. The latter was determined using the microcalorimeter LKB 2107 as follows. The sample of definite mass was exposed to vacuum in the Butch type cell, thermostated, and then the excess amount of water was in troduced into the cell causing a forced filling of the material volume. The obtained results are presented in Table 5 and Fig. 4.

The following conclusions can be made on the basis of tile data presented in the Table 5:

1. Experimental value of $\mathbf{C^{\infty}_{H_2O}}$, is lower than the "theoretical" one. This may be explained by two causes: the decrease of the total volume (as in the case of "Algipore") and non-filling of a part of the material free volume by water.
2. A satisfactory correlation exists between the theoretical values of $\mathbf{C^{\infty}_{H_2O}}$ and $\Delta\mathbf{H}$. Thus, the main reason of the difference between experimental and theoretical values of $\mathbf{C^{\infty}_{H_2O}}$ is evidently the non-filling of a part of the material free volume by water.

3. The absence of correlation between maximal water sorption and wetting angle, defined on the external surfaces of the material, shows that the values, obtained as mentioned above, do not reflect real interaction of water with internal surface of collagen.

Table 4. Experimental and Theoretical Data of the Maximum Sorption of Water by Burn Dressings

Covering name (material)	$C^{\infty}_{H_2O}$, g/g		ρ_0, g/cm³
	Experimental	Theoretical	
Helitrex (collagen)	32 ± 2	33 ± 2	0.030 ± 0.007
Helitrex (collagen sponge)	58 ± 3	55 ± 3	0.018 ± 0.005
Collagen dressing	1.8 ± 0.1	2.8 ± 0.3	0.350 ± 0.07
Corretium-2 (collagen)	3.5 ± 0.3	3.3 ± 0.3	0.300 ± 0.07
Corretium-3 (collagen)	2.1 ± 0.2	3.0 ± 0.1	0.330 ± 0.05
Combutec-2 (collagen)	77.0 ± 5.0	66.0 ± 3.0	0.015 ± 0.005
Epigard (foamy polyurethane	10.0 ± 0.3	15.0 ± 1.0	0.067 ± 0.005
Silicon-nylon composite	7.5 ± 0.2	7.7 ± 0.5	0.130 ± 0.03
Syspurderm (foamy polyurethane)	6.2 ± 0.2	7.1 ± 0.5	0.140 ± 0.03
Syncrite (foamy polyurethane)	20.0 ± 2.0	22.0 ± 1.5	0.050 ± 0.01
Farmexplant (foamy polyurethane)	12.0 ± 0.5	15.6 ± 3.0	0.064 ± 0.007
Johnson-Johnson (cellulose)	11.4 ± 0.5	10.0 ± 1.5	0.100 ± 0.03
Blood stopping (cellulose)	15.7 ± 0.9	12.5 ± 0.7	0.080 ± 0.006
Tunneling (cellulose)	4.3 ± 0.2	5.0 ± 0.4	0.200 ± 0.005
Switin (cellulose)	18.0 ± 2.0	20.0 ± 1.0	0.050 ± 0.005
Metallized (cellulose-paper)	12.4 ± 0.7	10.0 ± 0.5	0.100 ± 0.04
Needle-perforated (cellulose-viscose)	28.0 ± 2.5	30.0 ± 2.0	0.033 ± 0.007
Viscose 100%	25.0 ± 2.0	30.0 ± 2.0	0.033 ± 0.007
70% of cotton + 30% of viscose	31.0 ± 3.0	33.3 ± 3.0	0.030 ± 0.007
50% of cotton + 50% of viscose	25.0 ± 2.0	28.5 ± 2.0	0.035 ± 0.007
30% of cotton + 70% of viscose	28.0 ± 2.0	27.7 ± 2.0	0.036 ± 0.007
Algipore (vegetable)	30.0 ± 3.0	90.0 ± 5.0	0.011 ± 0.0002

Table 5. Density, Maximal Water Sorption, Wetting Angle and Heat Effect of Sorption of Water by Different Collagens.

ρ_0, g/cm^3	$c^{\infty}_{H_2O}$, g/g		φ°	ΔH, cal/g
	Experimental	Theoretical		
0.011	74	91	170	34.6
0.016	53	62.5	70	25.4
0.013	49	77	90	30.2
0.013	47	77	110	31.9
0.013	8	77	120	31.2
0.014	4	71.4	110	29.8
0.014	30	71.4	50	27.2

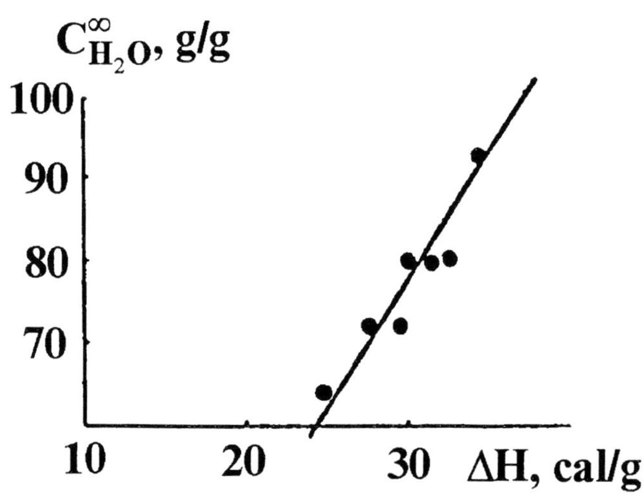

Figure 4. The dependence of maximum sorption of water by collagen materials on heat effect sorption.

Thus, it may be concluded that for the most number of burn dressings from hydrophilic materials, the maximum sorptional capacity with reference to water may be predicted satisfactorily. To fulfill this it is sufficient to use equation (6). For example, the experimental values of $C^{\infty}_{H_2O}$, correlate well with the free volume part of the materials (Fig. 5). The correlation coefficient is 0.96.

Maximal Sorption of Plasma by Burn Dressings

Sorption of the blood plasma by burn dressings was determined by a similar method. Plasma was obtained by centrifugation of the conserved blood. The treatment of the experimental results was carried out similarly to the case of the investigation of maximal sorption of water. The value of C^{∞}_{plasma}, differs from $C^{\infty}_{H_2O}$. The difference is not higher than 10%, that is why the data for C^{∞}_{plasma} are not shown in Table 5.

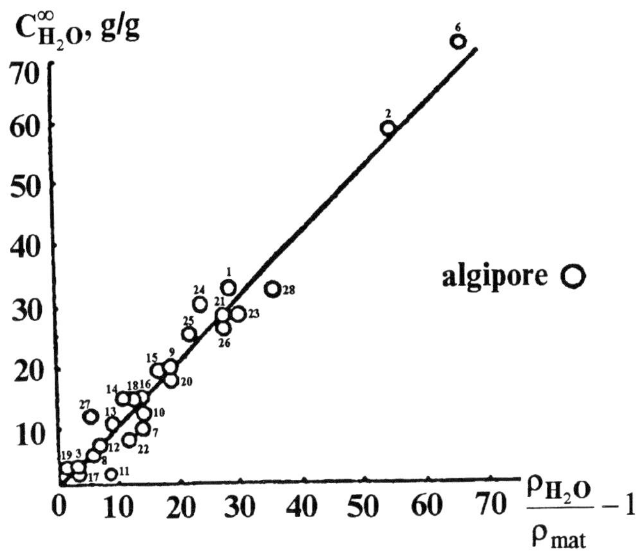

Figure 5. The dependence of $C^{\infty}_{H_2O}$ on free volume of the material various burn dressings: 1 - collagen dressing Helitrex; 2 - collagen sponge neutron; 3 - collagen dressing Braun; 4, 5 - artificial leather Corretium 2 and 3; 6 - Combutek-11; 7 - synthetic dressing Epigard; 8 - foamy polyurethane dressing Syspurderm; 9 - synthetic dressing Syncrite; 10 - foamy polyurohane dressing Famlexplant; 1 - burn face mask; 12 - compositional dressing Biobrant; 13 - cellulose dressing Johnson-Johnson; 14 - cellulose dressing Candall; 15 - cellulose non-adhesive dressing Torcatee; 16 - blood-stopping cellulose dressing: 17 - cotton balling dressing Mesorb; 18 - dressing with tunneling effect; 19 - cellulose dressing with nonadhesive synthetic layer; 20 - cotton balling dressing Svutyn; 21, 22 - metallized dressings; 23, 24 - needle perforated fabric with atraumatic layer; 25 - viscose 100%; 26-29 - viscose + cotton balling.

5.2 Study of the Kinetics of the Sorption of Liquid Media by Burn Dressings

The study of the kinetics of the sorption of the wound exudate by burn dressings is of great importance for the estimation of their efficiency.

The difficulties occurs at the mathematical description of the kinetics of the sorption process, connected with the absence of strictly quantitative description of the dressings structure.

Structure of Burn Dressings

Burn dressings are heterogeneous systems, consisting of several component-phases. As the general attention in dressings must be paid to the material possessing the maximum penetrability with reference to liquid medium, it is necessary to classify the types of heterogeneous systems.

For example, the penetrable parts of the material are placed under the layer of another weakly penetrable material in the way that diffusing flow is perpendicular to the surface layer. This is the case of double-layered dressings with a dense external layer. The penetrable parts of the material can be dispersed in a continuous weakly penetrable phase.

Dressings, based on collagen and cellulose, possess fibrillar structure and the fibers are randomly placed. In some cases spatial orientation of fibers is present. The sufficient amount of open pores in the dressings of such type is large but the open pores possess irregular form and great tortuosity in the direction of mass transfer. Modern burn dressings are multilayered with more dense external layer. Table 6.6 shows the mean-radius of macropores and their amount per unit of square for dressings, based on polyurethane.

All dressings possess a mean-radius of macropores in the range of $(2.0 \div 3.0)\times10^{-2}$ cm and a sufficiently narrow distribution (Fig. 6).

Theoretical

The detailed analysis of a number of mathematical models and results of experimental investigations of heterogeneous systems was performed by Barrer [45].

Table 6. Mean-Radius of Macropores and Their Amount Per Unit of Square N for Dressings, Based on Polyurethane.

Dressing name	R, cm×10^{-2}	N, pore/cm^2
Epigard	2.2 ± 0.2	370 ± 10
Syspurderm	1.8 ± 0.2	266 ± 5
Syncrite	2.8 ± 0.2	275 ±5
Farmexplant	2.2 ± 0.2	300 ± 10

Dressings are of a membrane form. If a membrane is in contact with the solution in a way that the concentration at one of its surfaces equals $\mathbf{C_0}$ and at the other equal 0 at $\mathbf{t} = 0$, then the total amount of the substance entered the membrane during time $\mathbf{t}(\mathbf{m_t})$, is given by the equation:

$$\frac{\mathbf{m_\infty}}{\mathbf{m_t}} = 1 - \frac{8}{\pi^2}\exp\left[-\frac{\mathbf{D}\pi^2\mathbf{t}}{\mathbf{l}^2}\right], \qquad (8)$$

where $\mathbf{m_\infty}$ is the amount of substance, entered the membrane at $t \to \infty$, i.e. in the equilibrium; **D** is the coefficient of substance diffusion in the membrane; **l** is the membrane thickness.

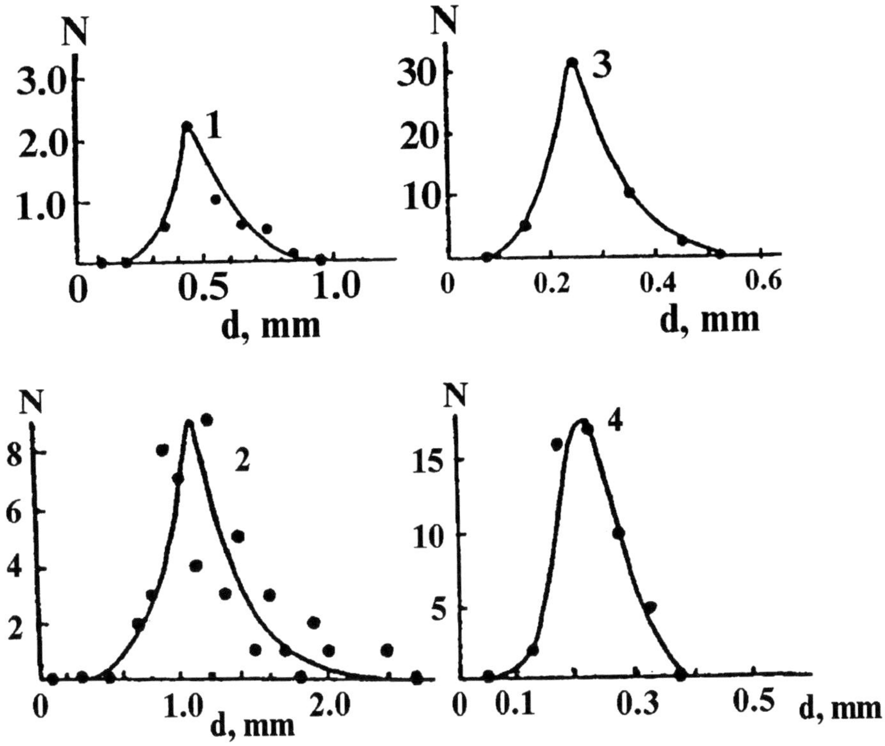

Figure 6. Curves of distribution of pores by sizes for different dressings: 1 -Farmexplant; 2 - Syncrite; 3 - collagen dressing Braun; 4 - Syspurderm.

The following correlation is satisfactory for the initial part of the kinetic curve of sorption:

$$\frac{m_t}{m_\infty} = 2\left[\frac{Dt}{\pi l^2}\right]^{1/2}. \tag{9}$$

Equation (8) was obtained for homogeneous material assuming that **D** does not depend on concentration of the substance in membrane. From equation (8) it is possible to calculate **D** value and time of membrane saturation by the substance up to a definite limit. For example, time of saturation up to $\frac{m_t}{m_\infty} = 0.85$ is calculated from the correlation:

$$\tau_{0.85} = \frac{l^2}{6D}, \tag{10}$$

and $\frac{m_t}{m_\infty} = 0.97$ is reached during the time

$$\tau_{0.97} = \frac{l^2}{3D}, \tag{11}$$

Both these correlations may be applied for practical calculations.

As it was mentioned above, the calculation of diffusion coefficient in heterogeneous systems is very difficult.

According to ideas accepted in the present time, the penetration of liquid into porous body is ruled by the laws of capillarity. These ideas are successfully applied for interpreting the penetration of water into paper, leather, fabrics, etc. [46].

Capillary pressure, which is the driving force of liquid to rise, is determined from the Jurene equation [47]:

$$P_k = \frac{2\gamma_i \cos\theta}{r}, \tag{12}$$

where γ_i is the surface tension of liquid; θ is the wetting angle; r is the capillary radius.

The equation, taking into account real structure of porous body, was obtained by Deriagin [47].

Experimental

Kinetics of sorption of water and blood plasma was investigated using the device for the maximal sorption of water.

Fig. 7 shows typical kinetic curves of sorption of water and plasma by various dressings. All curves are satisfactorily described by equation (9). Table 7 shows the values of $\frac{D_{eff}}{l^2}$ and the value of $\tau_{0.85}$, calculated according to equation (10).

The following conclusions could be made from the Table 7 data:

1. Burn dressings differ significantly in their rates of sorption of liquid media.
2. The rate of sorption is determined by the pores size and the material hydrophilicity.

5.3 Determination of Vapor Penetrability of Burn Dressings

With multilayered dressings, the external layer of is more dense than lower one which regulates the mass transfer of water from the wound into the surrounding. The process of mass transfer of water through the material layer is often called aquapenetrability or vapor penetrability.

Penetrability and diffusion of water in polymers were the subject of numerous investigations, the results of which are generalized in the list of reviews and monographs [43, 49] and are presented in Table 8. The mass transfer of water molecules in polymers possess a list of features. In hydrophilic matrices the interaction between water molecules and the material matrix is weak (low solubility). Nevertheless, the interaction of water molecules with each other stipulate a specific transfer mechanism.

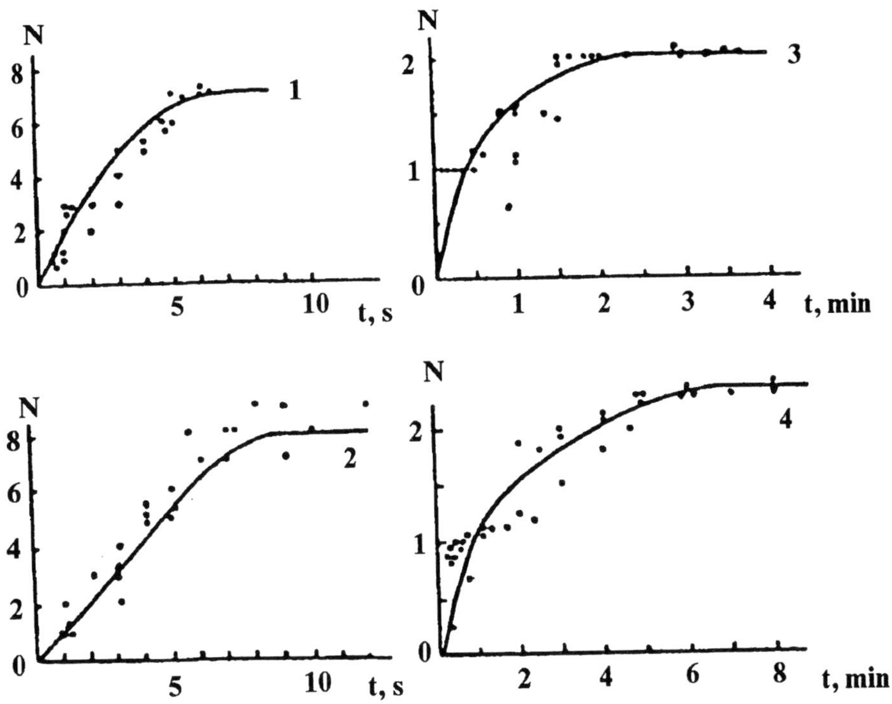

Figure 7. Curves of solution of water and blood plasma by different burn dressings: 1 - water - by needle perforated material; 2 - plasma - by needle perforated material; 3 - water - by polyurethane dressing Syspurderm; 4 - plasma - by Syspurderm dressing.

Table 7. Values of $\frac{D_{eff}}{l^2}$ and $\tau_{0.85}$ for Different Burn Dressings in Water andPlasma at 37°C.

Dressing name (material)	$\frac{D_{eff}}{l^2}\cdot 10^3$, s		$\tau_{0.85}\cdot 10^{-3}$, s	
	Water	Plasma	Water	Plasma
Helitrex (collagen)	2.5 ± 0.3	1.3 ± 0.2	3.9 ± 0.2	7.0 ± 0.3
Helitrex (collagen sponge)	6.6 ± 0.6	3.3 ± 0.4	1.5 ± 0.1	3.0 ± 0.15
Corretium-2 (collagen)	4.7 ± 0.4	2.5 ± 0.3	4.1 ± 0.2	8.0 ± 0.4
Combutek (collagen)	7.0 ± 0.7	3.5 ± 0.4	1.4 ± 0.1	3.0 ± 0.15
Syspurderm (foamy polyurethane)	2.3 ± 0.3	1.1 ± 0.1	4.3 ± 0.2	8.6 ± 0.4
Syncrite (foamy polyurethane)	(9.8 ± 0.8)*	(5.0 ± 0.5)*	90 ± 10	200 ± 10
Switin (cellulose)	(4.5 ± 0.4)*	(2.2 ± 0.2)*	210 ± 20	420 ± 35
Needle-perforated (cellulose)	35 ± 3	17 ± 1.7	0.3 ± 0.02	0.6 ± 0.05

* Must be multiplied by 10^{-2}.

Table 8. Penetrability and Diffusion of Water Vapors in Polymers [49].

Polymer	T, K	$\frac{p}{p_0}$	P·10[15], mol·m/m[2]·s·Pa	D·10[12], m[2]/s
Cellulose	298	1.0	8500	—
Regenerated cellulose	298	0.2	5700	0.1
Cellulose acetate	303	0.5 - 1.0	2000	1.7
Cellulose diacetate	298	1.0	15.7	—
Cellulose triacetate	298	1.0	5.5	—
Ethylcellulose	298	0.84	7950	18
Polydimethylorganosiloxane	308	0.2	14400	7000
Polyethylene (= 0.922)	298	0 - 0.1	30	23
Polyethyleneterephthalate	298	0 - 0.1	58.6	0.39
Polypropylene	298	0 - 0.1	17	24
Polyvinyl chloride	303	—	—	2.3
Polycaproamide	298	0.5	134	0.097

In hydrophilic materials the interaction between water molecules and hydrophilic groups of the material matrix stipulates high solubility of water in the matrix and increased aquapenetrability.

Thus, high aquapenetrability may be the property of hydrophobic as well as of hydrophilic materials, however the causes will be different. For example, in hydrophilic polydimethylorgano-siloxane the high mobility of water molecules is stipulated by high mobility of the chain units in this polymer. That is why despite low solubilities of water in polydimethylorganosiloxane the coefficient of aqua penetrability is significant.

On the opposite, in regenerated cellulose the diffusion coefficient is low because only the dissolved water molecules which are not connected with the matrix of this polymer participate in the mass transfer. In this case high value of aquapenetrability is stipulated by increasing the dissolved water content in regenerative cellulose which increases the part of water molecules participating in the mass transfer. This in turn leads to the increase of both diffusion coefficient and penetrability coefficient.

Theoretical

The mass transfer of water through a porous body is practically equal to that of gases in a polymer, provided there is no interaction between water molecules and the matrix of polymeric material.

Since the hydrophilic materials are commonly used for the production of dressings, which actively interact with water molecules, the diffusion should be considered simultaneously with absorption.

As a rule, the rate of the absorption process is significantly higher than the diffusion rate. Therefore, it can be assumed that the absorption equilibrium is immediately reached, and the concentration of water in the material C_{H_2O} is obtained from the following equation:

$$\frac{\partial C_{H_2O}}{\partial t} = D_{H_2O}\frac{\partial^2 C_{H_2O}}{\partial x^2} - \frac{\partial C^a_{H_2O}}{\partial t}, \tag{13}$$

where D_{H_2O} is the coefficient of water diffusion in the material; **x** is the diffusion coordinate; $C^a_{H_2O}$ is the concentration of the absorbed water.

The concentration of the absorbed water can be calculated for the particular cases. For example, if the concentration of functional groups, capable to link water molecules irreversibly, is limited and equals C_f, we can assume that the bonded water molecules no longer participate in the diffusion process, but form domains on which fast absorption occurs.

For the case when the concentration of water on one of the surfaces (**x** = 0) is constant and equals $C^0_{H_2O}$, the reaction zone reaches the second surface of the membrane, which is l thick, during the time **t** [50].

Thus, during the time **t** there will be no water flow through the surface **X** = 1 on the membrane exit, and then stationary flow will be set immediately. The amount of water passed through the membrane will equal

$$m_{H_2O} = D_{H_2O}\frac{\Delta C_{H_2O}}{l}\cdot S\cdot t, \tag{14}$$

where **S** is the square of the membrane; $\frac{\Delta C_{H_2O}}{l}$ is the concentration gradient. If the solubility of water in the material is ruled by the Henry law:

$$C_{H_2O} = \sigma P, \tag{15}$$

where **P** is the pressure of water vapors over the material; then substituting (15) into (14) we obtain the following equation:

$$m_{H_2O} = D_{H_2O}\cdot\sigma_{H_2O}\frac{\Delta P}{l}St. \tag{16}$$

Considering the diffusional coefficient D_{H_2O} being equal:

$$P_{H_2O} = D_{H_2O}\cdot\sigma_{H_2O}, \tag{17}$$

we obtain

$$P_{H_2O} = \frac{m_{H_2O}}{\Delta P \cdot S \cdot t}. \tag{18}$$

Experimental

Aquapenetrability of burn dressings was determined on the device, described in Section 3. The values of penetrability coefficients were calculated by Equation (18).

Table 9 shows the values of coefficients of aquapenetrability P_{H_2O} for various burn dressings.

Table 9. Values of Aquapenetrability Coefficients of Burn Dressings at 37°C.

Dressing name (material)	$P_{H_2O} \cdot 10^9$, mol·m/m²·s·Pa
Helitrex (collagen)	1.6 ± 0.1
Helitrex sponge (collagen)	11.0 ± 1.0
Brown dressing (collagen)	6.6 ± 0.6
Syspurderm (foamy polyurethane)	0.8 ± 0.2
Syncrite (foamy polyurethane)	1.2 ± 0.2
Epigard (foamy polyurethane)	4.3 ± 0.4
Farmexplant (foamy polyurethane)	3.3 ± 0.3
Biobrant (silicon-polyamide)	1.6 ± 0.16
Johnson-Johnson (cellulose)	2.0 ± 0.2
Perforated metallized dressing (cellulose)	9.0 ± 0.7
Face mask dressing (cellulose)	2.5 ± 0.2
Burn towel (cellulose)	5.4 ± 0.5
50% cotton + 50% viscose	8.0 ± 0.7
70% cotton + 30% viscose	8.0 ± 0.7
100% viscose	7.0 ± 0.7

5.4 Determination of the Air Penetrability of Burn Dressing

As mentioned in Section 5.1, active sorption of the wound exudate occurs during several minutes after the dressing closes the burn wound. Further on, there proceeds the evaporation of water from the external side of the dressing. This leads to the change of the state of exudate in the material mass. On the whole, this changes penetrability of the dressing with respect to air. In this case in order for anaerobic conditions not to be created in the wound, it is necessary to provide optimal air penetrability during the entire period of application.

The data on penetrability of dressings according to dry air are known in literature. Thus, for example, it is recommended [51] to determine air penetrability with the help of industrially produced VPTM-2 device. This device records automatically the amount of the air, passed through the dressing of the known square during time **t** at pressure oscillations of about 5 mm H_2O. However the application of such device does not allow us to investigate the air penetrability of dense materials such as foamy polyurethane compositions, and most importantly of dressings in wet state.

The construction and principle of action of the device, developed by the authors, allowing the determination of air penetrability of any material in any states and conditions, were described in Section 3.

Penetrability of Various Materials According to Oxygen and Nitrogen

The coefficient of gas penetrability (as well as of vapor penetrability) is calculated according to the formula (18).

The literature data on penetrability or various polymers according to the oxygen and nitrogen are given in Table 10.

Table 10.

Polymer	Penetrability coefficient, $(mol \cdot m/m^2 \cdot s \cdot Pa) \times 10^{15}$		Separation coefficient,
	O_2	N_2	O_2 - N_2
Polycaproamide	0.013	0.0033	3.8
Polyvinylchloride	0.022	0.008	2.8
Polyurethanic elastomer	0.032	0.10	3.2
Polyethylene ($\rho = 0.922$)	0.35	0.13	2.7
Polystyrene	3.13	0.73	2.9
Teflon	2.07	0.67	3.1
Ethylcellulose	3.2	0.93	3.4
Polydimethylsiloxane	168	83.0	2.0
Silicon rubber	200	87.0	2.3

As it is seen from the data presented in the Table 10, the penetrability of polymers may differ by four orders of magnitude. Special attention should be paid to high gas penetrability of polydimethylsiloxane and compositions on its basis, which is the result of the increased solubility of gases in them at high rate of diffusion (Table 11).

Table 11. Values of Penetrability Coefficients P ($mol \cdot m/m^2 \cdot s \cdot Pa$), Diffusion D ($m^2/s$) and Solubility σ ($mol/m^3 \cdot Pa$) of Gases into Polydimethylsiloxane at 20°C [52].

Gases	$P \cdot 10^{15}$	$D \cdot 10^{10}$	$\sigma \cdot 10^6$
O_2	83	23.3	36
N	164	30	55.6
CO_2	720	—	—

Penetrability of Porous Materials, Filled by Liquid Medium

A short list of studies considering the investigations of gas penetrability of polymeric membranes, in con tact with a liquid is given in [53].

It is observed that the sorption of liquid by a polymer leads to a decrease of gas penetrability coefficient in comparison with the liquid free polymer.

Theoretical

Let us consider the mass transfer of the air through porous body in two cases: one in which the free volume of all pores is filled by the air and the other with free volume filled by liquid medium. Porous body may be presented as consisting of two phases: the material forming the body's carcass and free space.

We also assume that pores possess cubic form and are disposed in the volume of the body, not joining each other. Such model is sufficiently suitable for porous burn dressings.

Let us determine the total thickness of the body in the direction of the mass transfer, total thickness of free space, occupied by pores, and total thickness of the layer, occupied by the material.

The total thickness of the body in the direction of the mass transfer is

$$\mathbf{l}_{\Sigma} = \frac{\mathbf{V}}{\mathbf{S}}, \tag{19}$$

where **V** and **S** are the volume and surface, respectively. Total thickness of the free space occupied by pores is

$$\mathbf{Q}_{\text{pores}} = \mathbf{l}_{\Sigma}\mathbf{Q}^{1/3} = \frac{\mathbf{V}}{\mathbf{S}} \cdot \mathbf{Q}^{1/3}, \tag{20}$$

where $\mathbf{Q}_{\text{pores}} = \frac{\mathbf{V}_{\text{pores}}}{\mathbf{V}}$ is porosity.

The total thickness of the layer, occupied by the material, is

$$\mathbf{l}_{\text{mat}} = \mathbf{l}_{\Sigma} - \mathbf{l}_{\text{pores}} = \frac{\mathbf{V}}{\mathbf{S}}\left(1 - \mathbf{Q}^{1/3}\right). \tag{21}$$

Thus the air passing through the porous body will overcome the resistance of two layers, each possessing its own penetrability coefficient with respect to air.

Total penetrability coefficient $\mathbf{P}_{\Sigma}$ or the porous body equals

$$\frac{1}{\mathbf{P}_{\Sigma}} = \frac{\mathbf{l}_{\text{pores}}}{\mathbf{l}_{\Sigma}\mathbf{P}_{\text{pores}}} + \frac{\mathbf{l}_{\text{mat}}}{\mathbf{l}_{\Sigma}\mathbf{P}_{\text{mat}}}, \tag{22}$$

where $\mathbf{P}_{\text{pores}}$ and $\mathbf{P}_{\text{mat}}$ are penetrability coefficients of the medium, presenting in pores; and the material, forming the body's carcass, respectively.

Let determine the ratio of penetrability coefficients of the porous body according to air, when its pores are filled by liquid and air.

$$\frac{P_{\Sigma(l)}}{P_{\Sigma(air)}} = \frac{\dfrac{P_{mat}}{P_{air}} + 1}{\xi \dfrac{P_{mat}}{P_l}}, \tag{23}$$

where

$$\xi = \frac{Q^{1/3}}{1 - Q^{1/3}}. \tag{24}$$

Values of P_{mat} are shown in Table 12.

Table 12. Values of the Coefficients of Penetrability, Diffusion and Solubility of the Oxygen in Air, Water, Plasma and Blood at 37°C (Dimensions as in Table 11).

Medium	P	D	σ
Air	$2.5 \cdot 10^{-9}$	$2.7 \cdot 10^{-5}$	$9.4 \cdot 10^{-5}$
Water	$7.4 \cdot 10^{-14}$	$3.0 \cdot 10^{-9}$*	$2.5 \cdot 10^{-5}$*
Plasma	—	$2.0 \cdot 10^{-9}$*	—
Blood	$1.4 \cdot 10^{-14}$	$1.4 \cdot 10^{-9}$*	$1.0 \cdot 10^{-5}$*

* The values were taken from [53].

Values of P_{air} and P_l can be estimated from the coefficients of diffusion and solubility of oxygen in air, water, plasma and blood at 37°C.

Values of penetrability coefficients of the oxygen in various media may be calculated according to the following expression:

$$P = D \cdot \sigma. \tag{25}$$

For any material $P_{air} \gg P_{mat}$, so we obtain more simple expression:

$$\frac{P_{\Sigma(air)}}{P_{\Sigma(l)}} \approx \xi \frac{P_{mat}}{P_l} + 1. \tag{26}$$

As for the majority of the dressings $\xi \gg 1$, and P_{mat} and P_l are of the same degree, the decrease of air penetrability of the dressing at pores filling by liquid must be then significant.

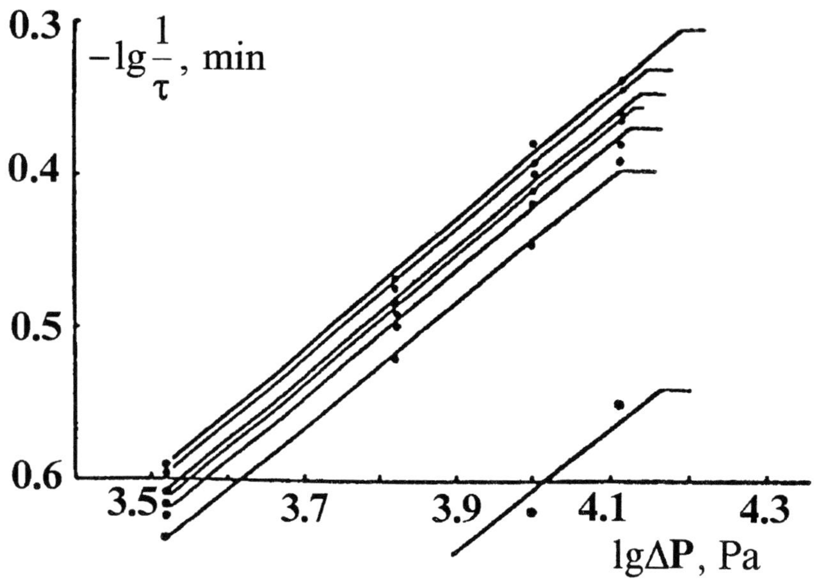

Figure 8. The dependence of $1/\lg(\tau)$ on pressure in the system for dry air: 1 - needle perforated material; 2 - collagen sponge; 3 - Syspurderm; 4 - Syncrite; 5 - cellulose dressing Svulyn; 6 - Farmexplant; 7 - Epigard.

Experimental

Air penetrability of burn dressings was determined using the de vice developed for this purpose (Section 4).

Two types of experiments were performed: determination of air penetrability for dry air and determination of air penetrability of dressings, preliminarily saturated by water (in conditions of maximal sorption of water), for humid air.

It has been shown by special experiment that air humidity (from 40 to 100%) does not practically influence the rate of penetration.

The experiments were performed according to the following scheme. At first we determined the time of filling by air of polyethylene sack of 45 1 volume in conditions, when the sample was not in the cell. This time (the constant of the device) depended on pressure in the system (**p**):

$$\lg\frac{1}{\mathbf{t}} = -2.00 + 0.44\lg\mathbf{p}. \quad (27)$$

The time of polyethylene sack filling at **p** = 100 Pa was selected as the standard. At **T**= (21 ± 1)°C, $\mathbf{t}_0 = 16.0 \pm 0.1$ min.

Subsequently the time of polyethylene sack filling with the sample was placed into the cell was similarly determined. It was observed (Fig. 8) that the dependence of $\mathbf{t}_x$ on **p** is described with the same slope as in equation (27) for all investigated dressings in conditions of the dry air penetration:

$$\lg\frac{1}{t_x} = -A_x + 0.44\lg p\,, \tag{28}$$

where A_x is the constant depending on structure and properties of the dressing material.

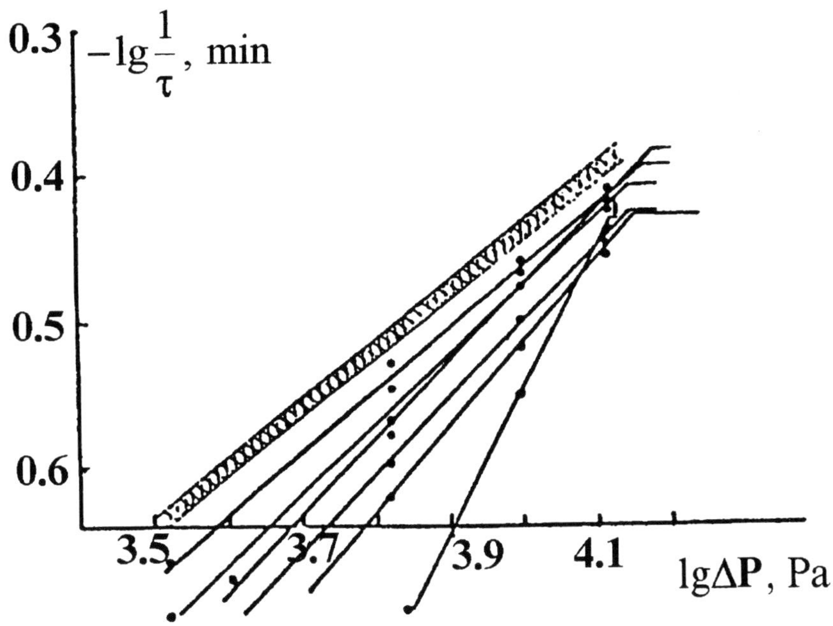

Figure 9. The dependence of 1/lg(τ) on pressure in the system for humid air: 1 – needle perforated material; 2 - Farmexplant; 3 - Combutek; 4 - Syspurderm; 5 - compositional dressing Biobrant (silicon-polyamide); 6 - Epigard.

At bubbling humid air through the dressing saturated by water, the slope increased significantly. That is why it is necessary to perform several experiments for each dressing at different pressures in order to extrapolate t_x to the pressure of 100 Pa with the required accuracy (Fig. 9).

The increase the slope of $\lg\left(\frac{1}{t_x}\right) - \lg p$ at air bubbling through the dressing saturated by water was attributed to the change of the material structure of the dressing resulting from the change of forms and sizes of macropores. This is often accompanied by a decrease of the total volume of the dressing.

Coefficient of air penetrability of the dressing (P_x) was calculated according to the equation:

$$P_x = \frac{ml_x}{S\cdot t\cdot\left(t_x - t_0\right)}, \tag{29}$$

where **m** is the polyethylene sack bulk equal 2 moles of air at (21 ± l)°C; S the surface square contacting with the bubbling air, equal $1.8 \cdot 10^{-3}$ m^2; **p** = 100 Pa.

Thus

$$\mathbf{P_x} = 11\frac{\mathbf{l_x}}{(\mathbf{t_x} - \mathbf{t_0})}. \tag{30}$$

The value of air penetrability coefficients for the dressings dry and saturated by water are shown in Table 13.

Table 13. Coefficients of Air Penetrability for Dry and Water-Saturated Burn Dressings at Temperature of (21 ± l)°C.

Dressing name (material)	Coefficient of air penetrability, mol·m/m²·s·Pa	
	Dry	Swollen
Helitrex (collagen)	$2.7 \cdot 10^{-5}$	0
Combutek (collagen)	$1.1 \cdot 10^{-3}$	0
Epigard (foamy polyurethane)	$1.3 \cdot 10^{-4}$	$1.3 \cdot 10^{-5}$
Syspurderm (foamy polyurethane)	$1.3 \cdot 10^{-4}$	$1.0 \cdot 10^{-6}$
Syncrite (foamy polyurethane)	$1.1 \cdot 10^{-3}$	$4.0 \cdot 10^{-5}$
Farmexplant (foamy polyurethane)	$4.5 \cdot 10^{-5}$	0
Biobrant (polyamide + silicon)	$1.8 \cdot 10^{-4}$	$7.0 \cdot 10^{-5}$
Johnson-Johnson (cellulose)	$1.6 \cdot 10^{-4}$	$3.0 \cdot 10^{-6}$
Needle-perforated material (cellulose)	$1.1 \cdot 10^{-3}$	—

From the data shown in Table 13 we see that on saturation with water a significant decrease of air penetrability takes place for all dressings, except for the "Biobrant".

Penetrability coefficient for dry dressings can be calculated according to the following equation:

$$\frac{1}{\mathbf{P}_{\Sigma(\mathrm{air})}} = \frac{\mathbf{Q}^{1/3}}{\mathbf{P}_{\mathrm{air}}} + \frac{(1+\mathbf{Q}^{1/3})}{\mathbf{P}_{\mathrm{mat}}}. \tag{31}$$

$\mathbf{P}_{air}$ is obtained from expression (25) using D = $2.7 \cdot 10^{-5}$ m^2/s and 45 mol/m^3 (solubility at the atmosphere pressure). The value of $\mathbf{P}_{air}$ is $1.2 \cdot 10^{-3}$ mol·m/m^2·s·Pa. The values of $\mathbf{P}_{\Sigma(air)}$ were taken from Table 13.

Values of $\mathbf{P}_{mat}$ were calculated from Equation (26).

The values of $\mathbf{P}_{\Sigma(H_2O)}$ can be obtained from the following equation:

$$\frac{1}{P_{\Sigma(H_2O)}} = \frac{Q^{1/3}}{P_{H_2O}} + \frac{\left(1+Q^{1/3}\right)}{P_{mat}}. \quad (32)$$

According to the calculations of $P_{\Sigma(H_2O)}$, its values fall close to 10^{-8} mole·m/m^2· s·Pa for the majority of dressings. It is this result which reveals the extremely low air penetrability for the listed dressings.

For some dressings the value of $P_{\Sigma(H_2O)}$ is significantly higher than 10^{-8} mole·m/m^2· s·Pa. This can be explained by two effects:

1. The presence of the air flow along the surface of pores (surface flow) [50];
2. The pressure of channels in the materials that are free of water.

To test these suppositions additional investigations are required.

5.5 Determination of Adhesion of Burn Dressings

Adhesion properties play a key role in the dressing performance. On one hand the lower layer of the dressing must be easily wetted, providing good adhesion of the dressing to the wound. On the other hand, the surface energy on the dressing - wound interface must be minimal to provide the smallest trauma on its removal from the wound.

Theoretical

The adhesive strength characterizes the ability of an adhesive structure to preserve its integrity. The adhesive strength as well as the strength of homogeneous solids is of kinetic nature. That is why the rate of tension increase and temperature affect the adhesive strength and the scale factor (i.e. sample dimensions) are also of great importance.

Different theories of adhesion of polymers were previously suggested [55].

1. Mechanical theory (MacBain), according to which the main role is devoted to mechanical filling of defects and pores of the surface (dressing) by the adhesive (blood).
2. Adsorptional theory (Mac-Loren) considering adhesion as a result of the performance of molecular interaction forces between contacting phases. According to this theory low adhesion, for example, may be reached between a substrate (dressing) with nonpolar groups and polar adhesive (blood).
3. Electrical theory (Deriagin) is based on the idea that the main factor controlling the strength of adhesive compounds rests in the double electrical layer which is formed on the adhesive-substrate interface.
4. Diffusional theory (Vojytzky) considers the adhesion as a result of interweaving of the polymers chains.

5. Molecular-kinetic theory (Lavrentiev) assumes that a continuous process of restoration and breakage of bonds proceeds in the zone of adhesive-substrate contact. Thus, the adhesive strength is defined by the difference between activation energies of the breakage and formation of bonds, and also depends on the correlation between the amount of segments participating in the formation of bonds and average amount of molecular bonds per unit of the contact area.

In recent years, the thermodynamic concept received the most attention. There, the main role is devoted to the correlation of surface energies of adhesive and substrate.

Thermodynamic work of adhesion of a liquid to a solid (**W**$_a$) is described by the Dupret-Jung equation:

$$\mathbf{W}_a = \gamma_l(1-\cos\theta), \tag{33}$$

where γ_l is the surface tension of liquid; θ is the wetting angle.

Substituting Jung's equation into equation (33)

$$\gamma_{s-l} = \gamma_s - \gamma_{s-l}\cos\theta, \tag{34}$$

we obtain the correlation

$$\mathbf{W}_a = \gamma_s + \gamma_l - \gamma_{s-l}, \tag{35}$$

where γ_s and γ_{s-l} are the surface tension of solid and of the solid-liquid interface, respectively.

It follows from equation (35) that the higher is **W**$_a$, the larger are the values of γ_s and γ_l while γ_{s-l} are smaller. However, according to (35) the increase of γ_s must, on one hand, lead to the growth of **W**$_a$, and, on the other hand, to an increase of γ_{s-l}. That is why the increase of the surface tension of the substrate is accompanied by the action of two effects. The necessary condition of the adhesive strength is $\gamma_l \gg \gamma_s$.

Values of γ_l and $\mathbf{W}_{s-H_2O}$ for different materials are shown in Table 14.

The value of γ_l for blood is 55.0 [56].

Experimental

Adhesive strength of burn dressings was determined according to the method, described in Section 3. Table 15 shows adhesive strength of various burn dressings and the angle of wetting by water also.

Table 14. Values of the Surface Tension and Thermodynamic Work of Adhesion of Various Materials [28].

Material	γ_s, mN/m	W_{s-H_2O}, mN/m
Polytetrafluoroethylene	18.5	83
Silicon rubber	21.0	78
Polyethylene	31.0	99
Polystyrene	33.0	105
Polymethylmethacrylate	39.0	103
Polyvinyl chloride	39.0	101
Polyethyleneterephthalate	43.0	104
Polycaproamide	46.0	107
Glass	170.0	222

Table 15. Adhesive Strength (a) and the Angle of Wetting by Water (θ) of Various Burn Dressings.

Dressing name (material)	A, mN/m	θ°
Corretium (collagen)	220 ± 20	75 ± 2
Syspurderm (foamy polyurethane)	210 ± 20	—
Epigard (foamy polyurethane)	350 ± 50	125 ± 3
Farmexplant (foamy polyurethane)	200 ± 20	130 ± 2
Bern-pack (cellulose)	170 ± 20	—
Biobrant (silicon-polyamide)	70 ± 10	—
Johnson-Johnson (cellulose)	20	—
Blood stopping non-adhesive dressing (cellulose)	20	—
Dressing with metallized lower layer (cellulose)	170 ± 50	—

6 The Model of a Burn Dressing Action

Three main processes proceed at the application of a dressing to the wound:

1. Sorption of the wound exudate by dressing.
2. Water evaporation from the dressing surface.
3. Mass transfer of gases through the dressing in conditions of the proceeding of sorption and evaporation processes.

Processes 1 and 3 were analyzed in detail in Section 4. It was found that sorption of the liquid media (water, plasma) proceeds fast and reaches the limiting value (maximal sorption ability) after several minutes for the most number of dressings, i.e. during the time sufficiently lower than that of the dressing action (2 - 3 days).

The mass transfer of gases (oxygen and nitrogen) through the dressing, is 2 - 4 orders of magnitude slower with wet samples, than with the dry ones in similar conditions.

Next, we consider the water evaporation from the dressing surface.

6.1 Evaporation of Water from the Dressing Surface

The dressing is saturated with water in air at 20°C temperature and 50% humidity. The temperature of the dressing surface is 32°C. This condition is chosen to take into account the temperature gradient in the matrix of the dressing.

Let us determine the amount of water, evaporating from the surface of the dressing during a given time period under stationery air, atmospheric pressure, and the dressing surface is completely saturated by water.

The partial pressure of air at 20°C and 50% relative humidity equals:

$$\mathbf{P_{H_2O}} = 1.26 \cdot 10^{-3}\ \mathrm{kg/cm^2},$$

$$\mathbf{P_{air}} = 1.02\ \mathrm{kg/cm^2}.$$

For air at 32°C in the saturation state:

$$\mathbf{P_{H_2O}} = 4.85 \cdot 10^{-2}\ \mathrm{kg/cm^2},$$

$$\mathbf{P_{air}} = 0.98\ \mathrm{kg/cm^2}.$$

The values for density, viscosity, heat conductivity and heat capacity of the air at the average temperature of 26°C equal:

$$\rho = 1.185\ \mathrm{kg/cm^3};$$

$$\mu = 1.861 \cdot 10^{-6}\ \mathrm{g/m \cdot s};$$

$$\lambda = 6.1 \cdot 10^{-6}\ \mathrm{kcal/m \cdot s \cdot deg};$$

$$C_p = 0.24\ \mathrm{kcal/k \cdot deg}.$$

After mathematical transformations using the method, described in [57-60], we may obtain the following equation for the mass transfer of water in dressing:

$$\mathbf{W} = \mathbf{a_m} \frac{\mathbf{P}}{\mathbf{R} \cdot \mathbf{T} \cdot \boldsymbol{\rho}_{\mathbf{av}}} (\mathbf{p}_1 - \mathbf{p}_2), \tag{36}$$

where $\mathbf{a_m}$ is the coefficient of heat conductivity; $\mathbf{p_{av}}$ is the average value of the mixture density over the surface and near the surface of the dressings; $\mathbf{p_1}$ and $\mathbf{p_2}$ is the partial pressure $\mathbf{p_{H_2O}}$ at 37°C and 20°C, respectively; **R** is the universal gas constant; **P** is normal pressure.

Substituting numerical values for the dressing of 1×1 m size, we obtain:

$$\mathbf{W} = 1.2 \cdot 10^{-1} \ \mathrm{g/m^2 s}.$$

If the dressing surface is not completely occupied by water, we should apply the following correlation:

$$\mathbf{w} = \frac{\mathbf{C}_{\mathrm{surf}(\mathbf{H_2O})}}{\mathbf{C}^0_{\mathrm{surf}(\mathbf{H_2O})}} \cdot 1.2 \cdot 10^{-1} \ \mathrm{g/m^2 s}, \tag{37}$$

where $\mathbf{C}_{\mathrm{surf}(\mathbf{H_2O})}$ and $\mathbf{C}^0_{\mathrm{surf}(\mathbf{H_2O})}$ are the surface concentration of water on the external side of the dressing and free water surface, respectively.

6.2 Sorption of Fluids by Burn Dressing from Bulk Containing a Definite Amount of Fluid

Let us consider the case, where the burn dressing is applied to a wound containing a definite amount of liquid. Assume that the dressing of a membrane of a given size (thickness and surface area **S**) is in contact with the solution of the restricted bulk **V**, which contains a $\mathbf{C_{0(s-s)}}$ concentration of diffusive substance. As the dressing is saturated by this substance, the concentration of the latter in the bulk will decrease.

The solution of the diffusional equation has the following form [57-60]:

$$\frac{\mathbf{m}}{\mathbf{m}_\infty} = 1 - \frac{2\mathbf{a}(1-\mathbf{a})}{1+\mathbf{a}+\mathbf{a}^2\mathbf{q}^2} \exp\left[\frac{4\mathbf{D}\mathbf{q}^2\mathbf{t}}{\mathbf{l}^2}\right], \tag{38}$$

where **q** is the positive solution of the characteristic equation

$$\mathrm{tg}\mathbf{q} = -\mathbf{a}\mathbf{q}; \ \mathbf{a} = \frac{\mathbf{V}}{\sigma \mathbf{S} \mathbf{l}},$$

where **σ** is the coefficient of distribution of the substance between the membrane and the solution.

When a sufficient part of the substance in solution is sorbed by the membrane, the value of **a** is small and a more simple expression can be used:

$$m_t \approx m^{\infty}\left[1-\frac{a}{\left(4\pi Dt/l^2\right)^{1/2}}\right]. \tag{39}$$

From Equations (38) and (39) we obtain two important correlations. Sorptional ability of the dressing, i.e. the part of the substance sorbed from the solution in the equilibrium conditions, equals:

$$\frac{m^{\infty}}{m^0}=\frac{1}{1+a}. \tag{40}$$

Thus, for the efficient action of the dressing it is necessary that the concentration **C** to be as high as possible in relation to the products of metabolism and toxins. Relating to water $C_{H_2O} \sim 1$, it is desirable the dressing volume (**l·S**) to be close to that of the exudate of wounds **V**.

The time of reaching 0.85 degree of maximum sorption of liquid media by the dressing equals:

$$\tau_{0.85}=12\frac{a^2 \cdot l^2}{\pi D}=12\frac{V^2}{\pi D\sigma^2 S^2}. \tag{41}$$

It depends on many parameters, each being able to affect in order the time of completion of the sorption process.

6.3 Mass Transfer of Water from Wound Info the Surrounding

Generally the change of the water amount under the dressing in the wound $\left(m_{H_2O}\right)$ is determined from the correlation derived from Equations (37) and (39):

$$\mathbf{m}_{(\mathbf{H_2O})} = \mathbf{V} \cdot \mathbf{C}^0_{\mathbf{H_2O}} - \mathbf{m}_{\mathbf{H_2O}(\text{dressing})}\left[1 - \frac{\mathbf{a}}{\left(4\pi\mathbf{Dt}/\mathbf{l}^2\right)^{1/2}}\right] - \tag{42}$$

$$-\frac{\mathbf{C}_{\text{surf}(\mathbf{H_2O})}}{\mathbf{C}^0_{\text{surf}(\mathbf{H_2O})}} \times 1.2 \cdot 10^{-1} \cdot \mathbf{S} \cdot \mathbf{t}.$$

Let consider the application of the correlation (42) for the following case. The wound characteristics are:

$\mathbf{S} = 10^{-2}\ \text{m}^{-2}$,

$\mathbf{C}^0_{\mathbf{H_2O}} = 10^6\ \text{g/m}^3$, $\mathbf{m}^0_{\mathbf{H_2O}} = 50\ \text{g}$, $\mathbf{V} = 5 \cdot 10^{-5}\ \text{m}^3$,

$\mathbf{S} = 10^{-2}\ \text{m}^{-2}$, $\mathbf{l} = 10^{-3}\ \text{m}$, $\boldsymbol{\sigma}_{\mathbf{H_2O}} = 1$,

$$\frac{\mathbf{C}_{\text{surf}(\mathbf{H_2O})}}{\mathbf{C}^0_{\text{surf}(\mathbf{H_2O})}} = 0.5.$$

For these conditions,

$$\tau_{0.85} = 12\frac{25 \cdot 10^{-10}}{\pi \cdot 10^{-9} \cdot 10^{-4}} = 9.5 \cdot 10^2\ \text{s (or} \sim 15\ \text{min)},$$

$$\frac{\mathbf{m}^\infty}{\mathbf{m}^0} = \frac{1}{1 + \dfrac{5 \cdot 10^{-5}}{10^{-2} \cdot 10^{-3}}} = 0.17 \text{ or } 8.5\ \text{g}.$$

During the same time the following amount of water will evaporate from the dressing surface:

$$\mathbf{m}_{\text{evap}(\mathbf{H_2O})} < 0.5 \cdot 1.2 \cdot 10^{-1} \cdot 950 \cdot 10^{-2} = 0.6\ \text{g},$$

i.e. the rate of evaporation is significantly (14 times) lower than that of water sorption by the dressing. All the amount of water from the wound (wound exudate) will evaporate during the time:

$$t = \frac{50}{0.5 \cdot 1.2 \cdot 10^{-1} \cdot 10^{-2}} = 8.3 \cdot 10^{4} \text{ s or} \sim 23 \text{ hours.}$$

7 Criteria of Efficiency of the First Aid Burn Dressings

7.1 The Requirements to the First Aid Dressings

The first aid burn dressing must meet the following criteria:

1. Sorption of the wound exudate, containing products of metabolism and toxic substances, during the period of the dressing action (24 - 48 hours).
2. Wound isolation from infection of the external medium.
3. Optimum air and water transfer between the wound and the surrounding.
4. Easy removal from the wound, causing no damage to the wound surface.

7.2 Characteristics of Burn Dressings

Below we list the characteristics of the burn wounds based on approximate estimations, discussed above. Note that no quantitative data are reported in the literature.

7.3 Rational Criteria for the Efficiency of the First Aid Burn Dressings

Sorptional Ability of Dressings

The burn wound (II - III degree) releases on the average $5 \cdot 10^3$ g/m^2 of the exudate. As it is seen from Table 3, the water amount is about 90%. The sorption of different components of the exudate proceeds with different rates. In this case, the free volume of the dressing material will be first filled with water. The diffusion of proteins and cells takes place in space occupied by water.

Modem burn dressings possess porosity of 0.9 (Table 4) and almost the entire free volume can be filled with water (Fig. 5). Maximum sorption ability for such dressings equals to:

$$C_{H_2O} \approx \frac{\rho_{H_2O}}{\rho},$$

and the amount of the liquid sorbed per square unit is:

$$\mathbf{m}_{\rho} \approx \mathbf{C}^{\infty}_{\mathbf{H_2O}} \cdot \rho \cdot \mathbf{l} = \frac{\rho_{\mathbf{H_2O}}}{\rho} \cdot \rho \cdot \mathbf{l} \approx 10^6 \cdot \mathbf{l}\ g/m^2$$

because $\rho_{\mathbf{H_2O}} = 10^6$ g/m^3.

As the burn dressing of the first aid must sorb $5 \cdot 10^3$ g/m^3, it follows that

$$5 \cdot 10^3 \approx 10^6 \cdot \mathbf{l}. \tag{43}$$

Therefore the thickness of tile first aid bum dressing equals

$$\mathbf{l} \approx \frac{5 \cdot 10^3}{10^6} \approx 5 \cdot 10^{-3}\ \mathrm{m}\ (0.5\ \mathrm{cm}).$$

Thus, the first criterion of the efficiency of the first aid burn dressing can be formulated as follows:

The burn dressing of the first aid must use its entire free volume for sorption. This volume must be 0.9 or more of the total volume of dressing. Dressing thickness must be 0.5 cm or more.

It should be mentioned that the majority of foreign first aid dressings fulfill this criterion.

Table 16. Characteristics of Burn Wounds [3].

Burn degree	Image of damage	Physiological process	Burn depth, mm
I	Redness and edema (medium edema)	Aseptic inflammatory process	0
II	Sac formation	Aseptic inflammatory process	0
III	Damage of skin cover, exudating wound surface	Skin necrosis, tissue necrosis	1 - 2
IV	Exudating wound surface	Full necrosis of tissues, carbonization of tissues	2 - 5

Table 17. Water Losses by Means of Evaporation from Burnt Surfaces of Different Types.

Surface type	Evaporation, ml/cm^2·hour
Natural skin	1 - 2
I-st burn degree	1 - 2.5
II-nd burn degree with sac intacts	2.8
II-nd degree with no damage of fermentative layer	37
II - IV degree of burn	20 - 31

Air Penetrability of Dressings

The air penetrability of dry air of the most of dressings is ranged within 10^{-4} - 10^{-5} mole·m/m^2·s·Pa (Table 13). The air penetrability of dressings filled with water is much lower and decreases to values between 10^{-6} - 10^{-5} mole·m/m^2·s·Pa, that is 0.2 - 2 dm^3/m^2·s.[1]

Thus, the second criterion of the efficiency of the first aid burn dressings can be formulated as follows:

The first aid burn dressing must possess air penetrability of 10^{-5} mole·m/m^2·s·Pa or higher after the sorption of water. For example, Biobrant burn dressing fulfills this criterion.

Adhesion of Dressing to Wound

Adhesion strength of dressings with respect to coagulated blood (Table 16) varies in a wide range, but it has the minimum value of ~ 20 N/m. This value should be accepted as the optimal one, because it corresponds to the minimal pain and damage on removal from the surface of natural skin. Thus, the third criterion of efficiency of the first aid burn dressings can be formulated as follows:

The first aid burn dressing must possess adhesive strength to the wound of 20 N/m or less after the end of its action.

The following burn dressings fulfill this criterion, for example: Biobrant, blood- stopping remedy, Johnson-Johnson.

Isolation of Wound from Infection from External Medium

It is known, that microorganisms, causing the wound infection, do not penetrate through the filters, possessing average pores size ~0.5 μm. The fourth criterion of efficiency of the first aid burn dressings is then as follows:

The first air burn dressings must possess no open pores with average diameter larger than $5 \cdot 10^{-7}$ m (0.5 μm).

Moreover, it is implied, that tile first aid burn dressings possess sufficient mechanical strength and elasticity both in dry and humid conditions.

8 Conclusion

The experimental methods to estimate the main physicochemical properties were worked out.

Based on theoretical and experimental data we found that maximal sorptional ability of burn dressing equals the free volume of the dressing material, calculated from the value of the material density.

We found that the study sorption ability water can be used as a model liquid instead of the blood plasma medium.

Kinetic parameters were determined from the sorption curves. These parameters showed that the first aid burn dressings markedly differ in the value of the rate or liquid media sorption at stages close to the sorption limits.

[1] This is for the dressing $5 \cdot 10^{-3}$ m (0.5 cm) thick at pressure of 50 Pa (5 mm H_2O) according to GOST 12088-77 (former USSR Standards).

We established that the air penetrability parameter in wet state decreases abruptly by 2 - 3 orders of magnitude for the majority of tested dressings. This is due to the filling of porous space by liquid medium.

We recommended that the air penetrability parameter be determined in wet state which represents the common condition of action for the first air burn dressing.

The value of adhesive strength after the end of its action on the wound must not exceed 20 N/m.

From the data obtained in this study we formulated the following criteria to estimate the efficiency of the fist aid burn dressing: maximum sorptional ability for water must be at least 10 g/g; optimal thickness of dressings, fulfilling this value of sorptional capacity, must be ~ $5 \cdot 10^{-3}$ m, the adhesive strength must not exceed 20 N/min, and the average diameter of open (connected) pores must not exceed $5 \cdot 10^{-7}$ m.

REFERENCES

[1] M.I. Fel'dshtein, V.S. Yakubovich, L.V. Raskina, T.T. Daurova, *Polymer Coatings for Wound and Burn Treatment*, Moscow, Institute of Information Publ., 1981, p. 299 (Rus)

[2] G.B. Park, *Biomaterials*, Med. Dev., Art. Org., 1978, **6**, 1.

[3] V. Rudkovsky, V. Nezelovsky, V. Zitkevich, N. Zinkevich, *Theory and Practice of Burn Treatment*, Moscow, Meditsina, 1988, p. 200. (Rus).

[4] A. Robin, K.H. Stenzel. In: *Biomaterials*, N.-Y., Plenum Press, 1969, 157-184.

[5] A. Robin, R.R. Riggio, R. Nachman, L.-*Trans. Soc. Art. Intern. Organs*, 1968, **14**, 1669.

[6] H.C. Grillo, I. Gross, *Surg. Res.*, 1962, **2**, 69.

[7] J. Oluwasanmi, M. Chapil, *J. Trauma*, 1976, **16**, 348.

[8] G. E. Zaikov, *Intern. J. Polym. Mater.*, 1994, **24**, 1.

[9] J.I. Abbendhaus, R.A. McMahon, J.G. Rosenkranz et al., *Surg. Forum*, 1965, **16**, 477.

[10] US Patent No. 35666871, 1971.

[11] FRG Patent. No. 1642112, 1973.

[12] US Patent No. 380078222, 1974.

[13] *Biomedical Business International*, 1981, **3**, 115.

[14] US Patent No. 3566871, 1971.

[15] US Patent No. 3521631, 1970.

[16] US Patent No. 3648692, 1972.

[17] UK Patent No. 1309768. 1973.

[18] G.L. Wilks, L.L.J. Samuels, *Biomed. Mater. Res.*, 1973, **7**, 541.

[19] I.A. Agureev, *Voenno-meditsinskii Zh.*, 1963, **6**, 74. (Rus)

[20] French Patent No. 2156068, 1973.

[21] K. Gorkisch, E. Vaubel, K. Hopf, *Comparative Clincial Studies on the Synthetic Wound Dressings*, Proc. II Intern. Congress on Plastics in Medicine, Netherlands, June, 1973.

[22] A.L. Iordanskii, G.E. Zaikov, T.E. Rudakova, *Interaction Between Polymers and Chemical and Biochemical Media*, Transport, Kinetics, Mechanism, Zeist (Utrecht), VSP Science Press, 1993, p. 288.

[23] USSR Certificate No. 245281, 1969, *Bulletin of Certificates*, No. 19.

[24] W.M. Chardack, M.M. Martin, T.C. Jewett et al., *Plast. Reconstruc. Surg.*, 1962, **30**, 554.
[25] C.W. Hall, D. Liotta, J.J. Chidoni et al., *J. Biomed. Mater. Res.*, 1972, **6**, 571.
[26] J.J. Guldarian. C. Jelenko, D. Calloway et al., *J. Trauma*, 1973, **13**, 32.
[27] US Patent No. 3875937, 1975.
[28] *Polymers for Medicine*, Ed. S. Madou, Moscow, Meditsina, 1981, p. 350 p. (Rus)
[29] USSR Certincate No. 267010, 1970, *Bullentin of Certificates*, No. 12.
[30] US Patent No. 358967, 1971.
[31] US Patent No. 3678933, 1972.
[32] US Patent No. 3654929, 1972.
[33] H. Kinkel, S. Holzman, *Chirurgie*, 1965, **36**, 535.
[34] USSR Certificate No. 6658148, 1979, *Bulletin of Certificates*, No. 15.
[35] M.I. Kuzin, V.K. Sologub, V.V. Yudenich el al., *Khirurgiya*, 1979, **8**, 86. (Rus)
[36] I.V. Yannas, J.F. Burke, *J. Biomed. Mater. Res.*, 1980, **14**, 65.
[37] S. Jacobson, U. Rothenaw, *J. Plast. Reconstr. Surg.*, 1976, **10**, 65.
[38] D. Spruit, K.E. Malten, *Dermat.*, 1966, **132**, 115.
[39] USSR Certificate No. 685292, 1979, *Bulletin of Certificates*, No. 34.
[40] *Textbook on Polymer Materials*, Moscow, Khilniya, 1980, p. 255. (Rus)
[41] G.E. Zaikov, A.L. Iordanskii, V.S. Markin, *Diffusion of Electrolytes in Polymers*, Utrecht, VSP Science Press, 1988, p. 328.
[42] Yu.V. Moiseev, G.E. Zaikov, *Chemical Resistance of Polymers in Reactive Media*, New York, Plenum Press, p. 586.
[43] I.A. Barrie, *Diffusion in Polymers*, London - New York, 1968, p. 452.
[44] S.I. Papkov, E.Z. Fainberg, *Interaction of Cellulose and Cellulose Materials with Water*, Moscow, Khimiya, 1976, p. 231. (Rus)
[45] R.M. Barrer, *Diffusion in Gases*, London - New York, Academic Press, 1968, p. 432.
[46] S.S. Voyutskii, *Physico-Chemical Principles of Fiber Materials Sorption by Polymers Dispersions*, Leningrad, Khimiya, 1969, p. 336. (Rus)
[47] D.A. Fridrikhsberg, *Course of Colloid Chemistry*, Leningrad, Khimiya, 1974, p. 351. (Rus)
[48] M.I. Al'tshuller, B.V. Deryagin, in: *Investigations in the Field of Surface Force*, Moscow, Nauka, 1967, p. 235. (Rus)
[49] M.M. Mikhailov, *Moisture Permeability of Organic Dielectrics*, Moscow - Leningrad, Gosenergoizdat, 1960, p. 162. (Rus)
[50] N.I. Nikolaev, *Diffusion in Membranes*, Moscow, Khimiya, 1980, p. 232. (Rus)
[51] *Textbook on Textile Materials*, Moscow, Legkaya Industriya, 1974, p. 342. (Rus)
[52] I.M. Raigorodskii,V.A. Savin, *Plasticheskie Massy*, 1976, **1**, 65. (Rus)
[53] V.N. Manin, A.N. Gromov, *Physico-Chemical Resistance of Polymer Materials During Exploitation*, Moscow, Khimiya, 1980, p. 247. (Rus)
[54] E. Laifut, *Transfer Phenomena in Living Systems*, Moscow, Mir, 1977, p. 520. (Rus)
[55] V.E. Basin, *Adhesion Durability*, Moscow, Khimiya, 1981, p. 208. (Rus)
[56] L. Sherstneva, A.E. Efremov, T. Ogorodova, *ISAO/IFAC Symposium Control Aspects of Artificial Organs. Transactions*, 1982, **4**, 845.

[57] L.M. Batuner, M.E. Pozin, *Mathematical Methods in Chemical Technology*, Leningrad, Khimiya, 1971, p. 822. (Rus)

[58] A. Jimenez, G. Zaikov, *Polymer Analysis and Degradation*, Nova Sci. Publ., New York, 2000, 287 p.

[59] G.E. Zaikov, *Polymers in Medicine*, Nova Sci. Publ., New York, 1998, 245 p.

[60] A.Ya. Polishchuk, G.E. Zaikov, *Multicomponent Transport in Polymer Systems*, Gordon and Breach, New York, 1996, 231 p.

Chapter 2

INFLUENCE OF MECHANICAL STRETCH ON SURFACE PROPERTIES OF OXIDIZED POLYPROPYLENE AND POLYETHYLENE

O. A. Legonkova, A. E. Chalykh and V. V. Ananiev
Moscow State University of Applied Biotechnology
Russia, 109316, Moscow, Talalihina str., 33

ABSTRACT

The work is devoted to investigation of energetic parameters of polyethylene and polypropylene surfaces under mechanical influence, during high temperature annealing and after corona and oxygen glow discharge treatment; to the influence of biaxial orientation on surface properties of polyolefines.
It is shown that the surface energy of treated and not treated with corona processing samples at different rate of stretching depends not only on temperature and degree of biaxial orientation, but also on contact of the surface with polar and no-polar mediums.
Treatment with corona discharge brings to emergence of oxygen–containing groups that don't vanish during temperature annealing, they simply re-direct from the surface into the bulk, remaining in the surface layer.

Key words: films, polyethylene, polypropylene, plasma of corona discharge, surface energy, polar and dispersive components of surface energy, biaxial orientation, temperature annealing.

INTRODUCTION

The current industry produces a great variety of polymer film materials. The most substantial part among these materials belongs to multilayer films with polyolefine layers. And there should be presence of certain complex of properties for successful use of multilayer films. In order to have high durable characteristics, the technology of their preparation usually includes the process of orientation. And it doesn't matter what method is used for getting multilayer films, the main aim is in receiving a good adhesion between layers. That's why the surface treatment (flame, different types of radiation, plasma of electrical discharge, etc.) is used.

So, surface treatment with a great variety of methods brings to better adhesion and to split resistance of films. Adhesion improvement is explained with emergence of electrical formations [1], disappearance of weak boundary layers [2], increasing of surface rough edges [3]. But the main reason of better adhesion between surfaces brought in touch is in chemical interaction [4-6].

Because of these reasons modification of surfaces still remains the most important stage in production of composite films' polymer materials. A significant dissemination has got the surface treatment in, so called, plasma of corona discharge, due to its manufacturability and rather low cost.

The influence of simultaneous biaxiall orientation at different rates of stretching, degree of orientation, temperature annealing, different mediums on behavior of plasma treated surface of PP and PE is the task of the present article.

EXPERIMENTAL

The following films were examined: polypropylene (PP), melting index 3.5 g/10min, and of polyethylene of low density (PE), melting index 2 g/10 min. The initial films with thickness 0.4 mm were received with the help of flat die extrusion on a chilled metal drum. Expansion was done simultaneously in two dimensions with the rate 1000 % per minute. Temperature in orientation chamber was 165 ^{0}C. After stretching heat fixing was carried out by withstanding of films at 155 ^{0}C for 3 min [7]. Surface of films was treated with plasma of corona discharge that has the following parameters: frequency - 50 Hz, gap between electrodes - 2 mm, voltage - 20 kV.

For measuring of polymers' surface energy, its polar and dispersive component, was used the method of determining of contact angles with a set of test liquids for wetted surfaces. [8]. As test liquids were used: water, glycerin, glycol, tricresyl phosphate, formamide, dimethylsulfoxide. For calculating of polar and dispersive component of surface energy we used equation of Owens and Wendt:

$$\frac{\gamma_{lv}(1+\cos\Theta)}{2(\gamma^{D}_{lv})^{1/2}} = (\gamma^{D}_{s})^{1/2} + (\gamma^{P}_{s})^{1/2}\left(\frac{\gamma^{P}_{lv}}{\gamma^{D}_{lv}}\right)^{1/2}$$

where Θ - contact angle of wetting, γ_{lv} – surface energy of wetting liquid, γ_{lv}^{D}- its dispersive; and γ_{lv}^{P} – polar components of surface energy; γ_{s} – surface energy of wetted surface, γ_{s}^{D} - its dispersive and γ_{s}^{P} – polar components.

The presence of oxygen containing groups in polymer was investigated through infrared spectroscopy on Specord M-80. The method of violated complete inner reflection was used to characterize the outer layer of polymer. Concentration of oxygen–containing groups was assessed through ratio of optical densities (D) at 1740 cm^{-1} (the region of absorbing of aliphatic aldehyde groups [9]) and 1460 cm^{-1} (intrinsic standard for PP and PE).

The following polymer samples were explored:

A) PP and PE without orientation; the same films treated with plasma of corona discharge;
B) PP simultaneously biaxially stretched at degrees of expansion: from 3 to 7; these films treated with plasma of corona discharge;
C) PP films, first treated in "corona" and then biaxially oriented at the same rates of stretching, i.e. from 3 to 7;

RESULTS AND DISCUSSIONS

The values of surface energy of initial (not treated) and treated with plasma of corona discharge PP films in dependence on the rate of biaxial stretching are given in the table 1. In this case previously oriented films were treated with plasma of corona discharge. In brackets the rate of biaxial stretching is given.

It is clearly seen from the table 1 that the surface energy of not treated with "corona" samples grows with the rate of stretching from 22.7 till 27 mJ\m^2. Polar component decreases. Thus, surface energy of initial PP films during simultaneous biaxial stretching grows only because of increasing of dispersive component.

Table 1. Surface Energy of Not Treated and Treated with Corona Discharge PP Films in Dependence on the Rate of Stretching during the Process of Orientation

Samples	Surface energy of not treated with corona films, mJ\m^2			Surface energy of treated with corona films, mJ\m^2		
	γ^P	γ^D	γ	γ^P	γ^D	γ
PP (1x1)	2.7	20.0	22.7	17.8	20.5	38.3
PP (3x3)	2.3	21.9	24.2	11.1	25.3	36.4
PP (4x4)	2.0	23.7	25.7	9.7	25.5	35.2
PP (5x5)	1.7	24.5	26.2	8.5	26.1	34.6
PP (6x6)	1.5	25.5	27.0	8.1	25.4	33.5

Treatment with "corona" brings to significant grow of polar component of surface energy from 2.7 till 17.8 mJ\m^2. But with the increasing of the rate of stretching the value of its accelerations becomes less.

The sequence of operations between corona treatment and orientation has an essential influence on the state of the surface occurred in the result of action of these processes. The dependence of energetic parameters of the surface on the rate of stretching at different sequence of operations is given in figures 1 and 2.

As it is shown in the figures, the surface energy of not treated with "corona" PP films increases with growing of simultaneous stretching. Refinement with corona discharge of not stretched films (the rate of orientation equals to 1) leads to significant growing of surface energy. But further biaxiall orientation gives essential decrease of the value of the surface energy. This effect is particularly noticed at small rates of stretching. And treatment with corona discharge of preliminary oriented films brings an increase of surface energy, but the magnitude of growth decrease with enlarging of stretching.

In all cases an increase of the rate of biaxial simultaneous stretching while orientation leads to decreasing of the polar component of the surface energy, that is, to lowering of oxygen containing groups on the surface. Moreover if films were treated with corona at first and then stretched, corona's processing practically doesn't change the polar component of films (curves 1 and 2 in the figure 2 for stretched films nearly coincide).

The growth of despersive component of treated with "corona" films in the process of stretching reveals itself to a less degree. This effect can be explained with partial destroying of ordered structures of treated with "corona" samples.

To define how the way of surface treatment influences the energetic characteristics, the surfaces of stretched and unstretched films were treated with oxygen plasma of glow discharge for 10 min. Dependence of energetic characteristics of these films on degree of biaxiall stretching is given in the figure 3.

As it is shown in the figures 1-3, the character of slopes doesn't depend on the way of surface treatment. The differences can be noticed in the values of energetic characterictics. The bigger magnitudes of surface energy during glow discharge may be explained by longer treatment (10 min.) then during "corona" treatment (1 min.). Behavior of films treated with different discharges while stretching is similar.

Energy characteristics of surfaces of all explored films are stable in time under normal conditions. For example, values of dispersive and polar components of surface energy of treated with corona discharge films don't change for 30 days of storage. The received data and the data [10, 11] where was marked stability of energy characteristics of PP and PE are nearly the same. However, unlike the authors [11] we haven't noticed the sharp decrease of surface energy during the first 4 days after treatment.

For further study of the process of biaxiall stretching (because orientation is carried out at $T=165^0C$, and it's necessary to heat samples before stretching) it was interesting to assess the contribution of thermal effect on surface energy. Behavior of functional groups on the surface of polymers in dependence on temperature changing was examined in some works [12]. They explored polystyrene, polydimetylsiloxane and cured phenol-formaldehyde, supervising the similar results.

In our case "corona" treated and not stretched PP and PE films were exposed at raised temperature for 3 minutes (high temparature annealing). The influence of temperature on surface energy of the films is exhibited in the figure 4.

As it is shown in the figure 4, with temperature increasing the polar component of surface energy decreases both for PP and for PE (for PE from 19.5 till 5.6 mJ/m^2 at $T = 165^0C$, for PP from 17.8 till 2.7 mJ/m^2 at $T=200^0C$), and the dispersive component grows a little (from 19.8 till 24.4 mJ/m^2 at $T = 165^0C$ for PE, and for PP - from 20.5 till 23.9 mJ/m^2 at $T=200^0C$).

Thus, the heating of treated with plasma of corona discharge PP films till temperature above melting point gives the value of polar component of surface energy the same as has the initial PP samples without any treatment (γ^P =2.7 mJ\m^2). As for PE samples, the heating above the melting point hasn't returned the value of polar component of surface energy to the magnitude, corresponding to the untreated sample (γ^P = 1.1 mJ\m^2). We explain this effect with cross-linking occurred after treatment with "corona" in the upper layers that restrains the mobility of macromoleculars. The effect of long time preservation of high values of surface energy during storage after "corona" treatment can be explained with the same reason.

The dispersive component increases with temperature growing both for PP and PE films.

In order to find the reason of changings of polar and dispersive components of surface energy during annealing, we made calculation of activation energy of relaxation processes, the values of which are presented in the table 2. Activation energy (E^P) was calculated via changing of polar component and E^D – activation energy, calculated via changing of dispersive component of surface energy.

According to the literature [13] the activation energy of β - transitions is up to 37 kJ/mol. Thus, data, presented in the table 2, let us make a conclusion that "corona" treatment brings oxygen-containing groups in amorphous phase and changings in the values of polar and dispersive components of surface energy connect with rotating mobility of small segments of the main valency chain.

Table 2. Activation Energy of Relaxation Transition in PP and PE

Samples	E^P, kJ/mol	E^D, kJ/mol
PP	13.3	31.5
PE	36.4	36.9

So, one of the most interesting effect in the carried out investigations is decreasing of polar component of the surface energy during orientation.

If during annealing there is a diffusion of oxidized factions from the surface into the bulk, we will supervise less absorption in spectrums of violated complete inner reflection and increasing absorption of oxygen-containing groups in spectrums of infrared spectroscopy.

All spectrums of infrared spectroscopy were without changes for all samples. It means that neither processes of orientation, nor treatment with plasma of corona discharge don't change the concentration of oxygen-containing groups in the bulk of the polymer.

The results of analysis of violated complete inner reflection are presented in table 3.

As it is shown in the table 3, treatment of films with corona discharge brings an appearance of a noticeable amount of oxygen-containing groups in a very thin outer layer of polymers. During annealing the content of these groups near the surface doesn't change practically. Comparing these data with change of polar component of surface energy in treated with "corona" films during annealing, we can conclude the following: oxygen-containing groups appeared on the surface while corona discharge treatment don't disappear during annealing, they simply redirect inside polymer, staying at the same time in the outer surface layer.

Table 3. The Results of Violated Complete Inner Reflection

Material	The way of sample preparing	D_{1740}/D_{1460}
The initial, not stretched PP	(a) not treated with "corona"	0.28
	(b) treated with "corona"	0.55
	(c) treated with "corona" and annealed at T=165^{0}C for 3 min.	0.54
The initial, not stretched PE	(a) not treated with "corona"	0.09
	(b) treated with "corona"	0.12
	(c) treated with "corona" and annealed at T=165^{0}C for 3 min.	0.13
Stretched PP film, the degree of biaxial orientation 4x4	(a) not treated with "corona"	0.10
	(b) treated with "corona"	0,27
	(c) treated with "corona" and annealed at T=165^{0}C for 3 min.	0.23
Stretched PP film, the degree of biaxial orientation 5x5	(a) not treated with "corona	0.05
	(b) treated with "corona"	0.13
	(c) treated with "corona" and annealed at T=165^{0}C for 3 min.	0.13
Stretched PP film, the degree of biaxial orientation 6x6	(a) not treated with "corona	0.05
	(b) treated with "corona"	0.10
	(c) treated with "corona" and annealed at T=165^{0}C for 3 min.	0.09
Stretched PP film, the degree of biaxial orientation 7x7	(a) not treated with "corona	0.05
	(b) treated with "corona"	0.10
	(c) treated with "corona" and annealed at T=165^{0}C for 3 min.	0.08

In the experiments described above all processes, i.e. simultaneously biaxial stretching and annealing, were made when the surface of polymer contacted with no polar medium – the air. That's why it was interesting to find out what would happen with oxygen-containing groups during annealing in contact with medium, which surface consists of polar groups as well. As such type of medium we took polyurethane two-component glue, which is usually used for glueing polymers.

Results of the following experiment are presented in the figure 5, where initial, not oriented films of PP were chosen, stage 1. Stage 2 – surfaces of films a) and b) were treated with "corona". Stage 3 - a layer of glue was applied on the surface treated with corona discharge (b) and not treated with "corona" (c). Annealing of all films was done at T=165^{0}C during 3 min. After that the glue was moved away with the help of solvent (ethyl-acetate) and, stage 4, annealing was done once again to be sure that there are no traces of the glue. After each process energy characteristics of the surface were calculated. For comparison a test film (a) was annealed in the air.

As it is seen from the fig.5, films, not treated with plasma of corona discharge, but having been in contact with polar medium, have a value of polar component of surface energy a little bit more after annealing then at the beginning of the experiment. The magnitude of polar component of surface energy of treated with "corona" films that were in contact with polar medium decreased much less then the same value for films which were in contact with air.

Thus, the received results let us make the consequence that polar groups may both direct into the bulk of polymer in case of contact of a surface with not polar medium and to the upper layer in case of contact of a surface with polar medium.

CONCLUSIONS

1. Biaxial simultaneous orientation of treated with "corona" films substantially changes energetic characteristics of the surface, moreover a polar component changes even at small degrees of stretching. A dispersive component of surface energy increases in the process of orientation.
2. Changings in the values of polar and dispersive components of surface energy connect with rotating mobility of small segments of the main valency chain.
3. A change of polar component during treatment with "corona" is connected with appearance of oxygen-containing groups on polymer surface Polar groups redirect into the depths of the polymer only if the surface contacts with no-polar medium. If the medium is polar, oxygen-containing groups remain on the surface of the polymer.
4. Treatment of films with corona or glow discharge gives the same character of changes of surface energy parameters with the rate of stretching: a polar component of surface energy decreases and dispersive component increases.
5. It was confirmed that energy characteristics of surfaces of all explored films are stable in time for 30 days.

REFERENCES

[1] Schonhorn, H. And Ryan F.W., *J.Appl.Polymer Science*, 18, 235 (1974).

[2] Gerenser, L.J., Elman, J.F., Mason, M.G., et al., *Polymer* 26, 1162 (1985).

[3] Kim, C.Y., Evans, J., Goring, D.A., J., *J.Appl.Polymer Science*, 15, 1357 (1971).

[4] Dillard J.G., Spinu I.M. An Investigation of the Effect of Plasma Treatment on the Surface Properties and Adhesion in Sheet Molded Composite (SMC) Material, J.Adhesion, 1990, vol.31, pp. 137-159.1. Brewis DM. Adhesion to Polymers – How Important are Weak Boundary-Layers, *Int.J.of Adhesion and Adhesives*, 1993, 13 (4), pp. 251-256.

[5] Gilman A. B., Potapov V.C. *Applied Physics*. 1995, # 3-4, pp.14-22.

[6] Bichler C. H., Langowski H.C., Moosheimer U, Seifert B. Adhesion mechanism of aluminum oxide and silicon oxide on biaxially oriented polypropylene, PET, poly(vinyl chloride), *J. of Adhesion Sci. And Techn.*, 1997, 11 (2), pp. 233-246.

[7] Ananjev V.V., Dvoretskaya N.M., Degtjariova S.N., Legonkova O.A. *Receiving and investigation of properties of oriented polypropylene films*, "Packaging yesterday, today, tomorrow", March – April 2000, p.20-21.

[8] A.J.Kinlock, *Adhesion and Adhesives*, translation from English under edition of doctor of science.L.E.Pritykin, Moscow, «Mir», 1991, 484c.

[9] R.Silverstain, G. Bassler, T. Morril, *Specro-metric identification of organic compounds*, Moscow, "Mir", 1977, 590c.

[10] Yaoguang Yao, Xeushu Liu, Yufen Zhu. The influence of the Surface Cross-linking of the Hihg Density Polyethylene Induced by Plasma Treatment on the Durability of Treatment Efficiency, *Journal of Applied Polymer Science*, Vol. 48, 57-65 (1993).

[11] Massines F, Gouda G. A comparison of polypropylene-surface treatment by filamentary, homogeneous and glow discharges in helium at atmospheric pressure, *J. Phys. D: Appl. Phys.*, 1998, 31, 3411-3420.

[12] *Polymers' mixtures*, under ed.of D.Pol, S.Newman, v.1, 2, Moscow, Mir, 540c.

[13] Bartenev G.M., Barteneva A.G. *Relaxation properties of polymers*. M., Chemistry, 1992, p.300.

Chapter 3

THE MECHANISM OF STEREOREGULATION IN DIENE POLYMERIZATION ON LANTHANIDE CATALYTIC SYSTEMS

Yuri B. Monakov[1], Zufar M. Sabirov[1] and Guennadi E. Zaikov[2]
[1] Institute of Organic Chemistry, Ufa Research Center, Russian Academy of Sciences, pr.Oktyabrya 71, Ufa, 450054 Bashkortostan, Russia
[2] N.M. Emanuel Institute of Biochemical Physics, Russian Academy of Sciences 4 Kosygin Street, Moscow 117334, Russia

The main distinctive feature of lanthanide catalysts among others Ziegler-Natta catalysts is that they allow one to synthesize polydienes with a high content of *cis*-1,4-units. In most cases, lanthanides are used in the form of blends and concentrates. About 50% of lanthanides consumed worldwide are used for the production of catalysts for various chemical processes. The number of these processes can be increased by using lanthanide catalysts in the manufacture of synthetic rubbers. A great number of studies are devoted to polymerization of dienes with lanthanide catalytic systems. Many of these studies are concerned with the problems related to the mechanism that controls the microstructure of polydienes. Although not all aspects of stereoregulation have been clarified, many problems were solved and possible explanations were offered for some of others [1-4]

The objective of this work was to review the basis results of research of diene (butadiene, isoprene, and piperylene) polymerization with the $LnHal_3 \cdot nL\text{-}AlR_3$ (Ln =lanthanide, L = tributylphosphate, AlR_3 = triisobutylaluminum and diisobutylaluminum hydride) catalytic system and to analyze the role of such factors as the electronic and geometric structure of bimetallic active centers, anti-syn and π-σ-transitions of the terminal units of the growing polymer chains, and the nature of the lanthanide, diene, and organoaluminum component in the mechanism of stereoregulation.

PRINCIPAL GROUPS OF LANTHANIDE-BASED CATALYSTS

Thus far, the attention of researchers has mainly been attracted to the family of ion-coordination lanthanide catalysts, by which *cis*-polydienes can be synthesized. These catalysts can be classified into several groups.[3]

Catalytic systems of the type $LnHal_3 \cdot nL-AlR_3$ stands for lanthanide ($n \leq 3$) make up group

I. The initial compound of the lanthanide can be obtained by reacting its halide with the corresponding organic base.[5] The research is usually carried out with complexes containing various electron donor ligands L, such as alcohols[6,7], cyclic ethers[8], aliphatic esters of orthophosphoric acid[9-15], aliphatic, and cyclic sulfoxides.[12,14,16] Catalysts in which the initial lanthanide component LnX_3 does not contain a halogen make up group **II**. The catalysts of this type include octanoates[17,18], stearates[19-23], naphthenates[13,18], acetylacetonates of trivalent lanthanides[19], and salts of di-2-ethylhexyl phosphoric acid.[19,20,23] In addition to LnX_3 and AlR_3, these catalytic systems must include halogen-containing components (one of the most frequently used compounds is AlR_2Cl), the role of which is to halogenate the lanthanide by exchange reactions. Catalysts in which the initial halogen-containing compound of lanthanide has the Ln-C bond form group **III**. *cis*-Polymerization of dienes can be catalyzed, for example, by the system $R'LnHal_2$ nTHF-AlR_3[24-28]. The synthesis of lanthanide compounds of this type is usually performed in tetrahydrofurane by conducting the reaction of the oxidative addition of R'Hal (triphenylmethyl chloride, benzyl chloride, phenyl bromide, allyl iodide) to zero-valent lanthanides.[24,25,27,29] Chloride of oligoisoprene neodymium, which is not bound via solvation bonds to tetrahydrofuran (THF/Nd < 3), catalyzes *cis*-polymerization of dienes even when no organoaluminum cocatalyst is present.[30]

The use of the catalysts based on individual compounds of lanthanides of various structures may help to elucidate the nature of active centers. This explains the interest in the problems of synthesis and conversions of these compounds.[31-40]

FACTORS AFFECTING THE ACTIVITY AND STEREOSPECIFICITY OF CATALYSTS

Chemical Nature of Lanthanide

The activity of lanthanide-based coordination catalysts is strongly influenced by the chemical nature of the lanthanide component. Depending on the conditions of polymerization, the relative activities of lanthanides can vary slightly. In all cases, however, with regard to the rates of polymerization of dienes, lanthanides form the following series: La <Ce<Pr<Nd>Gd>Tb>Dy>Ho>Er.[11,14] In the reaction of polymerization of butadiene, the propagation rate constant changes as follows: for La - 10, for Ce - 40, for Nd -140, for Gd - 130 and for Dy - 30 l/(mol min).[41] A similar behavior is observed for *cis*-polymerization of isoprene[18] and piperylene[42]. Although the propagation rate constant depends on the chemical nature of lanthanide, in the case of polymerization of butadiene, the degree of stereospecificity is approximately the same for all most active catalysts.[41] For polymerization of isoprene, the content of 3,4-units in the *cis*-polymer decreases with a rise in the atomic number of the lanthanide (from 2.5 to 1%), whereas *trans*-1,4-units are not formed at all. The reaction of the addition of *cis*-1,4-units in polymerization of isoprene can occur via three different mechanisms. The data of ^{13}C NMR indicate that only one of these mechanisms, namely, "head-to-head" addition, is realized. This conclusion is also corroborated by the results of ozonolysis.[43] Thus, lanthanides catalysts can be used to synthesize polyisoprene with molecules almost identical to those of natural rubber. In the case of polymerization of *trans*-isomer of piperylene, the microstructure of the polymer formed strongly depends on the nature of the lanthanine used as a catalysts.[11]

It was found that for the lanthanides of Ce subgroup, the content of *cis*-1,4-units in polypiperylene increases with an increasing atomic number. In contrast, in the Y subgroup, the number of *cis*-1,4-units decreases with a decreasing atomic number. The most stereoregular polypiperylene is formed when the polymerization occurs on a gadolinium catalyst. For a neodymium catalyst, the polypiperylene chain consists 1,4- and 1,2-units. At the same time, a spectroscopic examination of polypiperylene synthesized on a neodymium catalyst gave no evidence of 3,4-units. The ozonolysis data show that 1,4-units in this polydiene form "head-to-tail" sequences.[22]

The Nature of the Ligand

It is known that the microstructure of polybutadiene synthesized with the use of catalytic systems on the basis of halides of elements with d-electron orbitals depends strongly on the nature of the halogen. A special feature of catalysts on the basis of f-elements (lanthanides, uranium) is that they do not exhibit this dependence.[44,45] Nevertheless, the effect of the chemical nature of halogen on the reactivity of active centers was observed for lanthanide-based catalysts. In the case of polymerization of butadiene, this was confirmed by the results of measurements of reaction rate constants.[14] For other types of polydienes, the influence of the nature of halogen can be seen from the microstructure of the polymers formed. In the series of catalysts containing chlorine, bromine, and iodine, the content of *cis*-1,4-units in polyisoprene slightly decreases, and the content of 3,4-units, correspondingly, increases.[14] The effect of nature of halogen is the strongest in polymerization of piperylene.[11]

Varying the ligand L in the initial component $LnHal_3{\cdot}3L$ does not change the microstructure of the polybutadiene formed.[14] Nevertheless, the analysis of propagation rate constants shows that active centers in catalytic systems on the basis of tributyl phosphate and diamyl sulfoxide complexes of the same halide of lanthanide are different.[14]

The correlation between the nature of the halogen and its presence in the catalyst, and the functioning of the active centers of the latter is very complicated and mainly depends on the nature of the ligands attached to the lanthanide. Polymerization catalyzed by compounds containing Ln-Hal bonds yields polydienes with a high content of *cis*-units. It was even thought that, in the absence of such bonds, lanthanide catalysts are not stereospecific.

Systems on the basis of polynuclear complexes $[(C_6F_5)_3GeI_7M_2]Ln$ 3L-$Al(i\text{-}C_4H_9)_3$, where M is either mercury or cadmium and Ln is either praseodium or neodymium (in the trivalent state), are catalytically active in the polymerization of dienes.[46] For instance, at 25°C, these systems catalyze a formation of polybutadiene with up to 90% of *cis*-1,4-units.

The products of ball-milling of lanthanide metals in aromatic hydrocarbons include active intermediates of carbene type [=CH-CH=CH-CH=CH-CH-Ln-] and initiate *cis*-1,4-polymerization of butadiene and isoprene. It is assumed that the role of the electron acceptor ligand is played by the bulk of the metal or by its complex with a polyconjugated (polyacetylene) fragment. The latter is formed as a result of polymerization of the solvent.[40]

Polybutadiene and polyisoprene with 76-95% of *cis*-1,4-units can be obtained by using a combination of triisobutylaluminum with alcoholate or carboxylate of lanthanide chemically bonded to finely divided silica gel, on the surface of which compounds like $[SiO_2]$-O-$Nd(OR)_2$ are formed.[47]

When the lanthanide has no electron acceptor groups, polymerization yields *trans*-polydienes. For instance, systems including alcoholate or carboxylate of lanthanide in combination with triisobutylaluminum have low activity, and give macromolecules with the content of *trans*-1,4-units up to 85%.[48] Organometallic derivatives of LnR_3 type catalyze *trans* - polymerization of dienes.[30,33,36]

When butadiene and isoprene are polymerized on halogen-free ion-coordination catalytic systems on the basis of compounds of d-elements, polymers are obtained which contain *1,2*- and, *3.4*-units, respectively. Lanthanide-based catalysts ensure "head-to-tail" addition of monomers, i.e., formation of 1,4-units, even when the catalyst contains no halogen. These catalysts are characterized by high stereospecificity and, depending on the types of ligands bonded to the central atom, direct the polymerization reaction so that the resulting polymer contains a majority of either *cis*-1,4 or *trans*-1,4-units. The only known case of when a lanthanide-based catalysts gave polybutadiene and polyisoprene macromolecules, which contained a mixture of *cis*- and *trans*-sequences and a considerable amount of *1,2*- and *3,4*-units, was that of phenylcarbin neodymium.[31,33]

The Structure of Organometallic Component of Lanthanide Catalysts

The polymerization of dienes occurs when the lanthanide catalyst contains organometallic component (usually, trialkylaluminum or dialkylaluminum hydride). Systems $LnHal_3 \cdot nL$-AlR_2Hal do not reveal catalytic activity. Moreover, dialkylaluminum halide acts as catalytic poison when added to systems $LnHal_3 \cdot nL$-AlR_3.[49] A similar effect was observed when the content of dialkylaluminum halide in catalysts of the second group exceeded a certain optimum value.[19]

The rate of polymerization of dienes at 25°C strongly depends on the structure of the organoaluminum component of the catalyst. Increasing the polymerization temperature results in a leveling of the activities of such systems.[14] In addition, at a polymerization temperature of 25°C, the yield of polyisoprene on the catalysts of the type $LnCl_3$ 3TBP-AlR_3, where Ln is Pr or Nd, decreases in the series $Al(n\text{-}C_{10}H_{21})_3 > Al(n\text{-}C_6H_{13})_3 > Al(C_2H_5)_3$. Raising the temperature to 80°C reverses the order in this series. This may be explained by the decomposition of associates at a higher temperature. The result of this decomposition is that the cocatalytic activity of AlR_3 becomes consistent with its reactivity. Increasing the polymerization temperature removes the anomaly in the dependence of the yield of polydienes versus the Al:Nd ratio. This anomaly is observed when the polymerization is performed with organoaluminum cocatalysts that are able to form associates.[49] These facts show that only the monomeric form of AlR_3 can form active centers. The number of active centers in neodymium catalysts containing higher (6-8 carbons in the radical R) is about half that in catalysts with triisobutylaluminum.[41,50] It was somewhat unexpected that the propagation rate constant depended on the nature of AlR_3 for both butadiene and isoprene.[18,50]

That the reactivity of lanthanide active centers depends on the nature of AlR_3 follows also from the microstructure of polydienes. Varying AlR_3 component results in changes in the relative number of *cis*-1,4 and *trans*-1,4-units in polybutadiene. At the same time, the content of *1,2*-units remains the same (about 0.6% at 25°C). The only exception is a catalyst with diisobutylaluminum hydride, in which case, approximately a threefold increase in the content of *1,2*-units is observed. As the concentration of butadiene decreases (<0.5% mol/l) and the polymerization temperature increases (from 25 to 80°C) the dependence of the number of *cis*-1,4-units in the polymer on the structure of AlR_3 becomes more pronounced.[10]

When an organomagnesium compound $i\text{-}C_8H_{17}MgC_4H_9$ was used instead of the organoaluminum compound (triisobutylaluminum), all other conditions being the same, the activity of the catalyst markedly decreases.[51]

To achieve reasonable conversions, the polymerization with catalyst $NdCl_3 3TBP\text{-}i\text{-}C_8H_{17}MgC_4H_9$ was conducted for several days (compound $i\text{-}C_8H_{17}MgC_4H_9$ is catalytically inactive under these conditions). For this catalyst, the yield of the polymer depends on the ration between the starting components with a sharply pronounced maximum at Mg/Nd = 12÷13, wich is not characteristic of the lanthanide systems.

The fact that the replacement of trialkylaluminum by dialkylmagnesium in the cis-regulating halide-containing catalyst $NdCl_3{\cdot}3TBP\text{-}AlR_3$ leads to the inversion of the stereospecificity was also unexpected. At Mg/Nd =12÷13, a highly stereoregular trans-polybutadiene is formed. This polymer contains 95% of *trans*-1,4-units and 5% of 1,2-units at a maximum activity of the catalyst $NdCl_3{\cdot}3TBP\text{-}i\text{-}C_8H_{17}MgC_4H_9$. At the Mg/Nd ratio increases and goes beyond the above range, not only does the yield of polydiene drop but its micrpostructure also changes: the share of 1,2-units increases and, when toluene is used as a solvent, *cis*-1,4-units appear in the polymer. Therefore, polybutadiene of mixed structure is formed at Mg/Nd = 60 in this solvent (the amounts of *cis*-1,4-, *trans*-1,4-, and 1,2-units are 47, 44 and 9%, respectively).

Although the inversion of stereospecificity of the ion-coordination system under the action of the organometallic compound is not quite common, nevertheless, in our opinion, it is consistent with the regularities conventional for the lanthanide systems. The *trans*-regulating effect of the halide-containing system is apparently associated with the formation of a catalytically active halide-free alkyl derivative of neodymium of LnR_3 type by the reaction between the starting components of the catalyst because it is known that $RLnCl_2$ initiates the *cis*-polymerization of dienes, whereas R_2LnCl is catalytically inactive.[28] Complete replacement of chloride ions by hydrocarbon radicals in the reaction of f-metal chloride with organomagnesium compound was reported in works.[52,53] This assumption is also supported by the fact that the microstructures of polybutadienes formed with the LnR_3 [30,33,36] and $NdCl_3{\cdot}3TBP\text{-}i\text{-}C_8H_{17}MgC_4H_9$ catalysts are similar.

The Nature of the Solvent

In all cases, the rate of process is higher in aliphatic than in aromatic hydrocarbons. The data for n-heptane and toluene show that in these solvents, propagation rates are different (table 1). A similar effect was observed earlier[28] for polymerization of isoprene in n-hexane [K_p= 730 l/(mol min)] and in toluene [K_p=240 l/(mol min)] at 25°C on the catalytic system $(C_6H_5)_3CNdCl_2$ THF-Al(i-$C_4H_9)_3$. The reduction in the propagation rate constant under the effect of aromatic hydrocarbon is due to hydrocarbon and diene monomer competing for coordination at the active center. The effect of slowing down the polymerization reaction is due to the fact that aromatic hydrocarbon forms an arene complex with the lanthanide atom. The higher the π-electron-donating ability of the aromatic hydrocarbon, the lower the rate constant for polydiene chain propagation. Thus, the rate of polymerization of dienes decreases in the series benzene > toluene > xylene.[11,14,26,28]

Table 1. Polymerization of Dienes in Toluene and Heptane at 25° Cin the Presence of Catalytic System $NdCl_3 \cdot 3TBP$-Al(i-$C_4H_9)_3$.[3]

Monomer	Solvent	cis-1,4-units (%)	K_p, l/(mol min)	C_{ac}* (%)
Butadiene	Toluene	95	140	7
	Heptane	95	470	6
Isoprene	Toluene	95	80	9
	Heptane	95	120	8
Piperylene	Toluene	64	14	9
	Heptane	78	70	6

Note: * C_{ac} - concentration of active centers

THE STRUCTURE OF ACTIVE CENTERS AND MECHANISM OF STEREOREGULATION IN THE POLYMERIZATION OF BUTADIENE

The main parameters of diene polymerization with lanthanide-based catalytic systems are similar to those of polymerization with ion-coordinated catalysts on the basis of d-metals. This can be seen from the following facts: polymerization of dienes has an anionic coordinated character[18]: at polymerization temperatures from 20 to 25°C, the reaction is of first order with respect to the monomer and catalyst (this property is independent of the natures of catalyst and hydrocarbon solvent, the only exception to this rule being the system considered in work[18]) for most of catalysts studied;[18,21,26,28,41] and, the apparent activation energy of the reaction of polymerization of dienes is of the order of 33.5 kJ/mol.[20,41] For lanthanide catalysts, the concentration of active centers is somewhat higher than for conventional Ziegler-Natta catalysts, e.g., for neodimium-based catalysts their content varies from 6-10% [12,41,42,50] to 15-20%.[55-57]

It is generally accepted that, in the polymerization of dienes on lanthanide catalysts, the growing chain is attached to the transition metal by π-allyl bond and that the chain growth occurs by incorporation of the monomer via the metal-carbon σ-bond. In the case of neodymium catalysts, the delocalized π-allyl type structure of the terminal unit has been observed by spectroscopic methods.[8,26,28,58-60] The results reported in these works show that the relative contents of *cis*-1,4- and *trans*-1,4-units in polydienes depend on the type of solvent used, the polymerization temperature, structure of diene monomer, and the composition of lanthanide-based catalysts. These data can be interpreted in terms of the concept of isomerization equilibrium between *anti*- and *syn*-forms of π-allyl terminal unit. One of the arguments in favor of the existence of this of isomerization equilibrium is that the content of *cis*-1,4-units in polydienes decreases at the cost of increase in the content of *trans*-1,4-units when the concentration of the monomer is reduced.[10,28,54,61-64] More detailed information on the structure of the initial, terminal-, and penultimate unit of polydiene chains, and, correspondingly, information on the mechanism of the formation of chains, can be derived from the results of investigations of polymer structure by NMR. Works in this field are, however, not very numerous.[65]

With regard to the operation of lanthanide catalytical systems, one of the most important questions is that of the nature of the bond responsible for the growth of the polymer chain. Both the traditional Ziegler-Natta catalysts and lanthanide catalysts contain bridge bonds (of either chlorine, alkyl, or hydride type) between the metal atoms.[64-69] In view of this, it is usually assumed that the addition of molecules to the growing polymer chain occurs either via the metal-carbon σ-bonds (monometallic model of active center) or via the bridge bond (bimetallic model of active center). In the second model, the question arises as to whether the organoaluminum fragment constitutes a part of the active center or not. Direct experimental investigation of the structure of active sites is extremely difficult. Therefore, attempts have been made to analyze it indirectly. To this end, the effect of the structure of active centers on the kinetics of the corresponding reactions was investigated.[70-72] For polymerization of butadiene with $NdCl_3 \cdot 3TBP\text{-}AlR_3$ catalytic system, the reactivity of active centers and the dependence of the reactivity on the mechanism of introduction of AlR_3 were studied[70] using "precipitated" catalyst (precipitate of catalytic system multiple-wash with purified toluene). It was found that the reactivity K_p and the stereoregulating properties of active centers in the system $NdCl_3 \cdot 3TBP\text{-}AlR_3$ vary as the structure of AlR_3 is changed. The reactivity of these centers is similar to the reactivity of active centers initially prepared in the presence of the organoaluminum compound, which is added to the "precipitated" catalyst (table 2). This result (in combination with the data on the effect of the structure component of the catalyst on the microstructure of polydienes and the asymmetry of active centers and reactivity ratios[16,18,50]) suggests that the organoaluminum compounds constitutes a part of lanthanide active center and that, in the catalytic system, there occurs an exchange reaction, which results in the substitution of the initial organoaluminum fragment (ligand) of the active center by a new one.

Table 2. Reactivity of Active Centers and Microstructure of Polymer in Butadiene Polymerization with the Precipitated Catalyst with AlR_3' Additives (Starting System, $NdCl_3$·3TBP-butadiene-AlR_3; [$NdCl_3$·3TBP] = $1x10^{-3}$, [AlR_3'] = $1.5x10^{-2}$, [butadiene] = 1.5 mol/l, toluene, 25°C.

Starting catalytic system	Additional AlR_3'	K_p, l/(mol min)	cis-1,4- (%)	trans-1,4- (%)	1,2- (%)
starting $NdCl_3$·3TBP-butadiene-$Al(i\text{-}C_4H_9)_3$	-	220	94.3	5.1	0.6
	$Al(i\text{-}C_4H_9)_3$	210	94.5	4.8	0.7
precipitated $NdCl_3$·3TBP-$Al(i\text{-}C_4H_9)_3$	$Al(i\text{-}C_4H_9)_2H$	45	93.5	4.8	1.7
starting $NdCl_3$·3TBP-butadiene-$Al(i\text{-}C_4H_9)_2H$	-	50	93.4	4.8	1.8
	$Al(i\text{-}C_4H_9)_3$	200	94.3	4.9	0.8
precipitated $NdCl_3$·3TBP-$Al(i\text{-}C_4H_9)_2H$	$Al(i\text{-}C_4H_9)_2H$	50	93.5	5.0	1.5

The assumption about the bimetallic bridge structure of lanthanide catalytic systems was made in many works.[66-69] Nevertheless, the possibility must not be ruled out that active centers contain both types of bonds (π-allyl and σ-bridge). It is important that these bonds may differ dramatically in reactivities. In order to answer the question concerning the possible coexistence of two types of bonds, the polymerization of butadiene on catalytic systems $NdCl_3$ 3L-AlR_3 (where R is i-C_4H_9; Ln is Nd or Tb; L is tributyl phosphate) prepared in the presence of a small amount of butadiene and piperylene[71,72] was investigated (table 3). It was shown that the initial reactivity of active centers in $LnCl_3$ ·3L-butadiene-AlR_3 (system 1) and in $LnCl_3$ ·3L-piperylene-AlR_3 (system 2) in polymerization of butadiene is different. Addition prior to polymerization of a small amount of butadiene ([diene]:[neodimium] =10) to system 2 and of the same amount of piperylene to system 1, and aging the system for certain time, results in that the systems "forget", as the residence time increases, what dienes were present during the preparation (formation) period. The reactivities of active centers of the two systems interchange (table 4). These changes in the reactivities can be due to the fact that active centers have bimetallic structure and include two metal-carbon bond, Ln-C (π-allyl) and Ln-C-Al (σ-bridge), via which the reaction of insertion of the monomer into the chain occur. These bonds differ greatly in reactivity. The experimentally observed dependence of the reactivity of active centers of catalytic systems on the residence time can be explained by two reasons: (1) slow rate of chain growth in the case when the propagation occurs via the least reactive bond, and (2) equilibrium reaction (e.g., migration of organoaluminum fragment in the active center) in the result of which the bridge bond (Ln-Al-C) between the endgroup and the lanthanide converts to π-allyl bond and vice versa.[71,72]

Table 3. Dependence of K_p in Butadiene Polymerization on the Time of Aging for Catalytic Systems 3 and 4 (That is, after Adding the Second Portion of Diene).

Time of aging catalyst t, min	K_p, l/(mol min), system 3	K_p, l/(mol min), system 4
*	330(1)	220(2)
12	290	240
30	230	
90		260
120	230	
180		280
240	220	
360	220	
1440	220	300

Notes: System 1: $NdCl_3$ 3L-piperylene-AlR_3; system 2: $NdCl_3$ 3L-butadiene-AlR_3; system 3: system 1 + butadiene; system4: system 2 + piperylene; [Al]/[Nd] = 30; [Nd] = 2, t_{prep} = 120 min, AlR_3 = $Al(i\text{-}C_4H_9)_3$, toluene, 25°C. * The values of rate constants for the starting systems 1 and 2.

We also note that some individual organometallic (or hydride) compounds of the type $RLnCl_2$ lead to the formation of *cis-* polydienes even when the organoaluminum component is absent. However, these catalysts are much less active than $LnCl_3 \cdot nL\text{-}AlR_3$.[30,38] Complexes of these compounds with THF do not catalyze polymerization of dienes. At the same time, the addition of AlR_3 gives an effective catalyst.[24-28] As evidenced by IR spectra, the addition of $Al(i\text{-}C_4H_9)_3$ to $(C_6H_5)_3CNdCl_2$ THF (Al:Nd = 3-5) results in that molecules of THF migrate from neodymium to the coordination sphere of aluminum atom.[26-28] In the reaction between $LnCl_3 \cdot nL$ and AlR_3, alkylation of lanthanide should occur prior to the shift of the ligand coordination bond from L to AlR_3, otherwise alkylation would be prohibited. It is of interest that the compounds of the type R_2LnCl do not exhibit catalytic activity. This property remains whether AlR_3 is present or not.[24,26,28]

Active centers of lanthanide catalyst probably differ in reactivity. This conclusion was made on the basis that polydienes obtained with lanthanide catalysts have a broad molecular mass distribution,[50,55] the polydispersity index being on the order of three even when the reactions of chain termination and transfer do not occur.[57]

Quantum-chemical calculations of the electronic structure of model active centers and analysis of their geometry was presented in works.[70,73-77] It was established that *anti-* conformation of model centers is less favorable (from the point of view of the total electron energy) than *syn-* conformation. The total energy of centers with σ-bonds is smaller than the total energy of centers with π-bonds. The fact that particular isomeric form of the active center exerts little influence on the electron density on the Ln-C_α bond (it is via this bond that the growth of polymer chain occurs) suggests that the reactivities of different isomeric forms of active centers are about the same. In the lanthanide series, the variation of the electronic structures of active centers is very small.

The calculation of structure of complexes formed by active center with monomer showed that, for all lanthanides, the complexes which include cisoid conformers of dienes are energetically more favorable. In the lanthanide series, the preference for complexes with cisoid conformers with respect to similar complexes with transoid conformers (we measure the preference in terms of the difference in the total energy of formation of corresponding complexes) changes from 4 to 7 kJ/mol. This means that, for cisoid conformers, the energy of complexation is from 19 to 23 kJ/mol larger than for transoid conformers.

The results of quantum-chemical calculation also show that isomerization transitions of active centers, *anti*-conformation-*syn*-conformation, are possible, in principle.[73-77]

If the insertion reaction occurs only in the σ-type complexes (i.e., only by the metal-carbon σ-bond), one should take into account the lifetime τ of those σ-form of the active center that allow insertion of the monomer. This time, which is determined by the kinetics of the processes occuring in the complex, should be compared to the characteristic time τ_1 of the elementary act of insertion. The role of the π-σ rearrangment in the mechanism of stereoregulation has been investigated works.[78,79] It was established that in cases where the terminal unit of the growing polymer chain has the form of a stable π-allylic structure (i.e., when the time spent by the terminal unit in σ- form is much shorter than the characteristic time of the elementary act of insertion), only the cisoid conformer of butadiene molecule can convert the terminal unit from π- to σ-form and then insert into the metal-carbon σ-bond. During the insertion, the active center remains in the σ-form. This result does not depend on the overall geometry of the active center (i.e., whether it is octahedral, tetrahedral, etc) and is explained exclusively by the presence of a stable, longlived π-allyl terminal unit. It has been assumed that it is the latter factor, i.e., the presence of a longlived π-terminal unit, that explains the fact that in some cases polydienes are formed with an exceptionally high content of cis-units.[70,79]

ROLE OF THE STRUCTURE OF DIENE IN THE MECHANISM OF REGIO- AND STEREOSELECTIVITY

The objective of this part of work was to study the role of the structure diene (butadiene, isoprene and piperylene) in the mechanism of regio- and stereoselectivity in polymerization with the lanthanide catalytic system $NdCl_3 \cdot 3TBP\text{-}Al(i\text{-}C_4H_9)_3$. To this end, we performed a quantum-chemical investigation of the electronic and geometric structures of the prereaction complexes of dienes (butadiene, piperylene, and isoprene) with an active center and of the complexes that model the transition state in the reaction of insertion of dienes via the Nd-C σ-bond in the active centers.[80] All calculations were performed using the quasi-relativistic extended Huckel method.[81]

It is known that, in diene polymerization with *cis*-regulating systems, the terminal units of the growing polymer chains have delocalized π-allylic structure.[8,68] High regioselectivity of the lanthanide catalysts in the polymerization of nonsymmetric dienes implies that the active centers of single type, in which the delocalized terminal unit is linked to the penultimate unit according to "head-to-tail" type, are predominantly formed during the polymerization. This is shown for (P-cis-$CH_2C_\gamma HC_\beta R'C_\alpha HR$-anti-$CH_2C_\gamma HC_\beta R'C_\alpha HR$-NdLL':

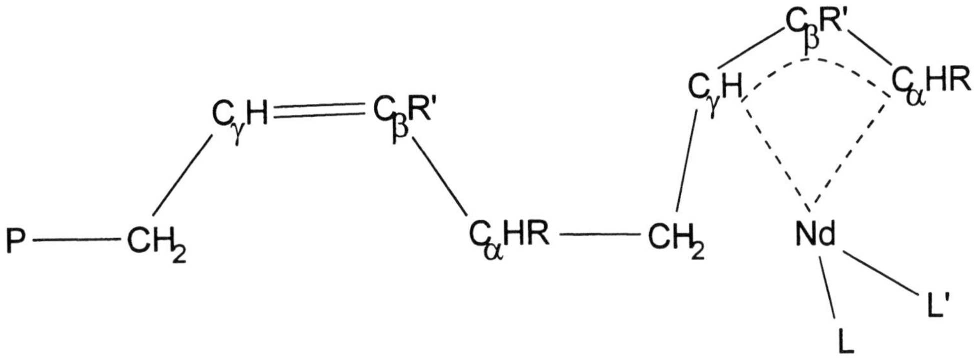

where P is the polymer chain, L and L' are the ligands determined by the initial composition of the catalytic system: R =CH_3 and R' = H in the polymerization of piperylene, R = H and R' = CH_3 in the polymerization of isoprene, and R = R' =H in polymerization of butadiene.

The active centers formed in the polymerization of dienes with the $NdCl_3 \cdot 3TBP$-Al(i-$C_4H_9)_3$ catalytic system were modeled by the compound that took into account the two terminal units of the polymer chain: $CH_2HC_\gamma HC_\beta R'C_\alpha HR$-$CH_2C_\gamma HC_\beta R'C_\alpha HR$-$NdCl_2 \cdot AlR_3$. The necessity of including the organoaluminum component in the composition of active centers was substantiated in works.[3,82]

Because the reaction of unsertion can proceed only via the metal-carbon σ-bond, at the first stage, the calculations were aimed at finding the manner in which the coordinating diene can stabilize the σ-structure of the active center. The calculations of prereaction complexes showed that both piperylene and isoprene (like butadiene[79]) can stabilize the σ-form of the center if they occupy the coordination site in the Nd sphere (fig.1c), which becomes vacant upon the transition of the active center from the π- to σ-structure (figs.1a and 1b). The energies of this complex formation are 7.1, 8.5, and 9.2 kJ/mol for butadiene, isoprene, and piperylene, respectively. At the second stage, while retaining the center in the σ-form, the diene can isomerize from the *transoid* to the *cisoid* conformation (due to the rotation of the-CH=CH_2 double bond about the C-C ordinary bond). The driving force of isomerization is that, in the *cisoid* conformation, the -C^3H=C^4H_2 bond of the diene reacts with the Nd-C_α σ-bond. This is the beginning of the insertion of the diene into the chain (fig.1d). The energy expended to accomplish the isomerization is small and is equal to ~12-15 kJ/mol, depending of the structure of diene. The complexes of *cisoid* isomers formed upon isomerization have the energies of complex formation equal to 8.4, 9.7, and 10.5 kJ/mol, for butadiene, isoprene, and piperylene, respectively. Consideration of the possible structures of prereaction complexes shows that when the diene retains the *transoid* conformation, because of geometric reasons it cannot stabilize the σ-structure of the active center and participate in the insertion at the same time.

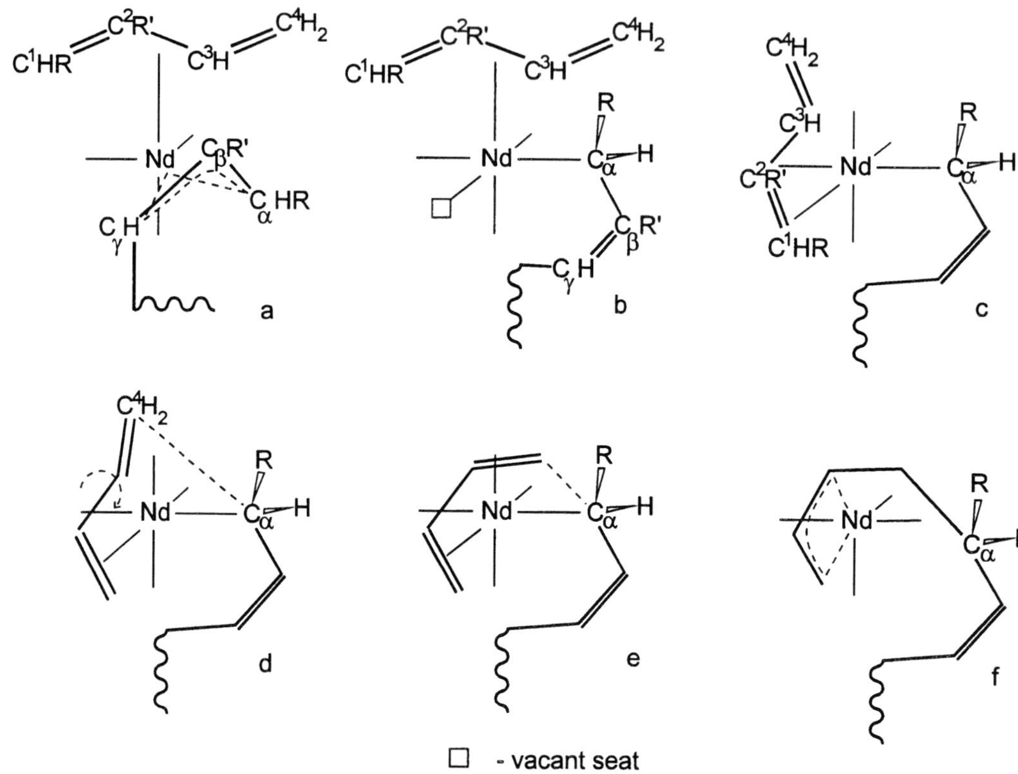

Figure 1. Basic stages of growing reaction by polymerization of butadiene (**R** = **R'** = **H**), isoprene (**R** = **H** and **R'** = **CH_3**) and piperylene (**R** = **CH_3** and **R'** = **H**) under the action of catalytic system **$NdCl_3 \cdot 3TBF$-Al (iC_4H_9)$_3$.**

Thus, *cis*-stereospecificity results from π-allylic stabilization of terminal unit of the growing polymer chain, when the lifetime of the terminal unit in the σ-state is not sufficient for the elementary event of insertion to take place. Otherwise, there is no need for σ-stabilization of the center by the formation of complex shown in fig.1c, and the dienes directly attack the Nd-C_α bond with one of its double bonds. Naturally, the dienes take the conformation that is thermodynamically more advantageous in solution, that is, the *transoid* conformation.

The occurence of the reaction of insertion in complex 1c necessitates that the diene further approaches the Nd atom; therefore, a certain energy is required to achieve the corresponding transition state. Taking into account that the structure of the active center in the polymerization of all dienes is similar and these complexes differ only by the presence and the arrangement of methyl substituents, the model structures of the transition states (fig.1e) were assumed to be similar. The calculations showed that the energy expended to form the transition states depends slightly on the structure of diene and is equal to 49, 54, and 57 kL/mol for butadiene, isoprene, and piperylene, respectively. The analysis of electronic populations of the bonds of the atoms of dienes and active centers (fig.1e, table 4) indicates that the electronic structure of the transition complexes are characterized by delocalization of the RHC^1=C^{2}R' double bond and by incorporation into the complex of the diene linked to the Nd atom by the three carbon atoms C^1, C^2, and C^3. The fourth carbon atom of the diene C^4 begins to form the bond (C^4-C$_\alpha$) with the terminal carbon atom of the growing polymer chain.

This implies that only a single double bond of the diene $C^3H=C^4H_2$ is involved in the insertion via the Nd-C_α bond. The second double bond $RHC^1=C^2R'$ reacts only with the Nd atom. This offers the possibility for the subsequent formation of a π-allylic bond with the Nd atom.

Table 4. Electron Bond Populations in Model Complexes cis-C_4H_7-syn-$C_4H_7NdCl_2$-$Al(CH_3)_3$ + C_4H_6

Complex on Fig.	Electron bond populations					
	Nd-C_1	Nd-C_2	Nd-C_3	Nd-C_4	Nd-C_α	C_α-C_4
1c	0.011	0.014	0.007	0.0	0.286	0.0
1d	0.043	0.047	0.030	0.008	0.096	0.235
1f	0.104	0.108	0.101	0.0	0.0	0.846

In fact, the calculation of complexes modeling the final state in the reaction of insertion (the length of the Nd-C_α bond is greater than 4.0A; the plane of the molecule of the inserted diene was brought closer to a distance of 2.6 A, which corresponds to the equilibrium distance of the plane of the crotyl terminal group to the Nd atom) showed that the π-allylic bond of the newly forming terminal unit with the Nd atom takes place concurrently with the insertion.

Table 4 shows that, in the final complex (fig. 1f), the Nd-C_α bond does not exit any longer, the new C^4-C_α bond is formed, and the C^3-C^4 bond is ordinary (the population is 0.85). All these facts suggest that the insertion took place. The second double bond is delocalized (the electron populations of C^1-C^2 and C^2-C^3 bonds are close to 1.01) and the three carbon atoms C^1, C^2, and C^3 form a π-allylic bond with the Nd atom.

As for the regioselectivity, it is important that in the presence of lanthanide catalysts the dienes can enter insertion (upon isomerization from the *transoid* to *cisoid* conformation) only via the double bond containing no methyl group. The reactions for this phenomenon are different for piperylene and isoprene.

In the polymerization of piperylene, when the $CH_2=CH(CH_3)$ doudle bond of the diene comes closer to the Nd-C_α σ-bond, the steric repulsion between the methyl groups of the terminal unit and the piperylene molecule appears (fig. 2a). Therefore, it is necessary to alter significantly the orientation of the terminal unit of the growing polymer chain, for example, to turn it about the Nd-C_α bond to decrease the repulsion of the methyl groups. Naturally, these changes are extended in time and need considerable expenditure of energy. In our opinion, it is this factor that is responsible for the participation of only the unsubstituted double bond of piperylene in the insertion and ensures the regioselectivity, that is, the addition of 1,4-units only according to the "head-to-tail" type.

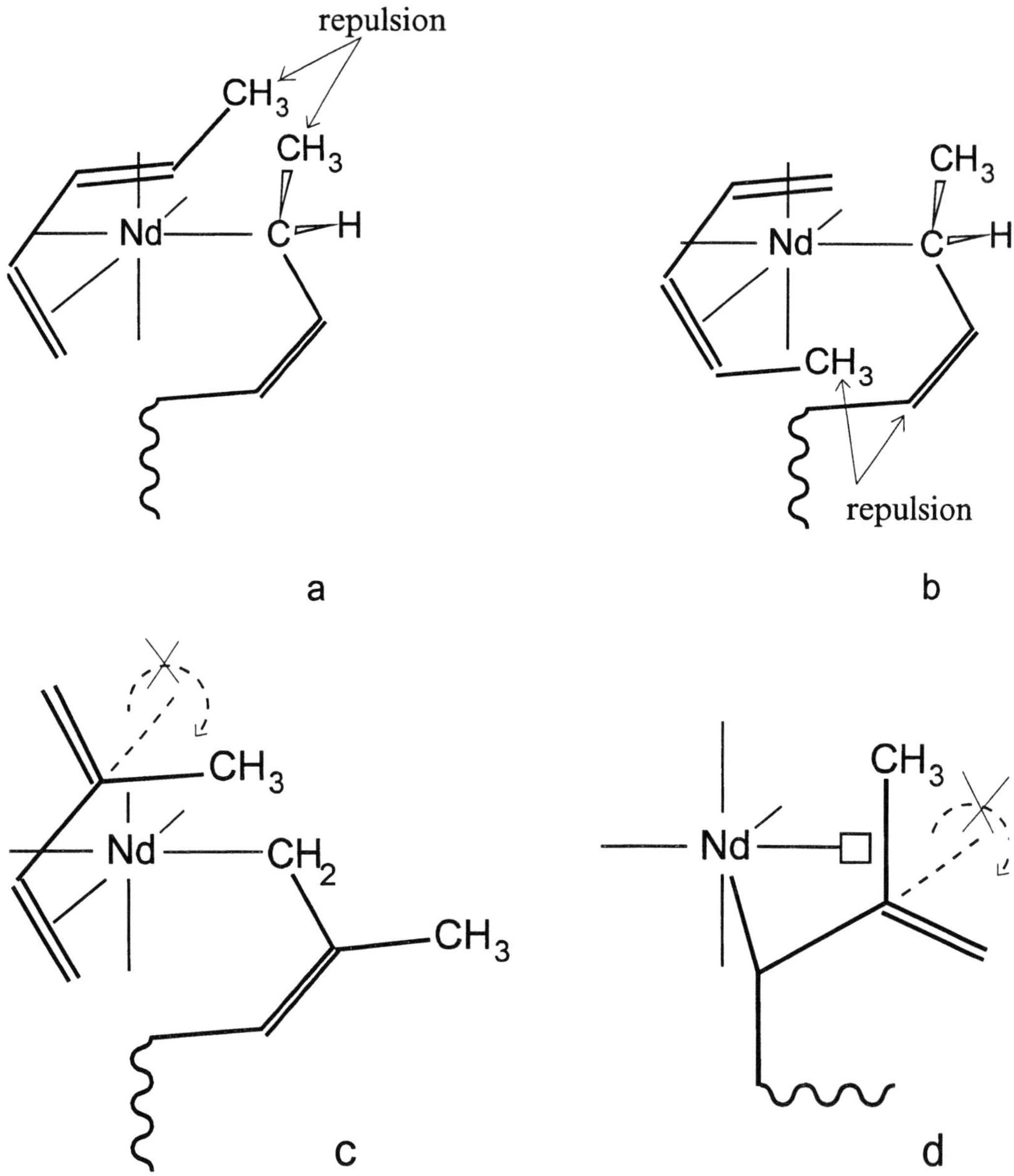

Figure 2. Appearance of steric hindrances by piperylene polymerization (a, b) and by isoprene polymerization (c, d).

As for the problem why the *cis*-isomers of piperylene are not polymerized with *cis*-regulating lanthanide catalysts, note that among the four possible isomers of piperylene (*cis-cisoid*, *cis-transoid*, *trans-cisoid*, and *trans-transoid*), *cis-cisoid* isomer is the least advantageous (according to the total electron energy) and this isomer apparently does not occur in the racemic mixtire in solution.[83] To take part in the reaction of insertion, piperylene must isomerize from the *transoid* to the *cisoid* conformation. In this case, the *cis-cisoid* conformation appears, which is the least advantageous for piperyene. Therefore, this isomerization may not take place. When the *cis-cisoid* isomer is involved in the insertion, the steric hindrances can appear, because at any orientation the methyl substituent turns out to be directed as if inwards into the center (for example, see fig.2b). All these facts predetermine the inactivity of the *cis*-isomer of piperylene in the polymerization with *cis*-regulating lanthanide catalysts. Naturally, in the polymerization initiated by the catalysts showing "mixed" stereospecificity or *trans*-stereospecificity, the transition of piperylene from the *transoid* to *cisoid* conformation is no larger necessary and, for example, the *cis-transoid* isomer can enter the polymerization.

Also note that, in accordance with the calculated data, the energies of the σ-form of the "piperylene" active center, in which the Nd atom is linked to the C_γ atom, and the σ-form of the center, in which the metal atom is bonded to the C_α atom, differ slightly. The corresponding value of ΔE is maximal, when the terminal units are in the *trans-cisoid* conformation and is as low as 4 kJ/mol. In the models of the "butadiene" centers, this difference is much greater (12.7 kJ/mol)[76], that is, in polymerization of piperylene, the σ-structure of the active center with the Nd-C_γ bond is realized more often than in the butadiene polymerization. This accounts for why the total amount of 1,2- and trans-1,4-units must be greater in the polymerization of piperylene than in the polymerization of butadiene (due to insertion via the Nd-C_γ bond and the *anti-syn* isomerization of the terminal unit). This way evidenced experimentally.[82]

In the polymerization of isoprene, the arrangement of diene in the prereaction complex (as shown in fig.2c) requires that the transition from the *transoid* to *cisoid* conformation takes place via the rotation of the $C(CH_3) = CH_2$ substituted bond. The fact that the methyl group and the double bond are aligned in the opposite directions with respect to the ordinary C^2-C^3 bond (contrary to piperylene, when they are situated on one side of the C^2-C^3 bond) hampers the isomerization. Upon rotation of the $-C(CH_3)=CH_2$ group by 180°, either the methyl or the $=CH_2$ group approaches the Nd atom too closely. This sharply increases the energy barrier (to 76 kJ/mol), in contrast to butadiene and piperylene, when the barrier to rotation is 11.9 and 14.9 kJ/mol, respectively. This explain why the double bond of isoprene, which bears a methyl substituent, cannot participate in the insertion. Therefore, regioselectivity appears and polyisoprene presumably does not contain the *trans*-units. In fact, upon insertion of isoprene, the active center is formed, whose σ-form (upon the formation of σ-bond with third carbon atom of the terminal unit) has a structure similar to that shown in fig.2d. It is seen that, as in the case of complex 2c, the rotation of the $-C(CH_3)=CH_2$ group is necessary for the isomerization of the terminal unit from *cis* to *trans* conformation to take place. It is the energy barrier of this isomerization that is responsible for the absence of *trans-1,4*-units in polyisoprene, in contrast to polypiperylene and polybutadiene prepared with the same *cis*-regulating catalysts.

Thus, the high stability of the terminal unit, which predominantly occurs in π-allylic state, leads to the participation of only the *cisoid* isomers of dienes in the reaction of insertion.

In the polymerization of piperylene regioselectivity (addition of 1,4-units according to "head-to-tail" type) is caused by the fact that only the double bond of diene bearing no methyl substituent can react with the metal-carbon σ-bond of the active center. Involvement of the double bond of piperylene bearing the methyl substituent in the reaction of insertion is considerably hampered because of the steric repulsion between the methyl groups of the terminal unit of the growing chain and the diene. The smaller difference in energy between the two possible σ-forms of the terminal unit in the polymerization of piperylene (as compared to butadiene) is responsible for the increased lifetime of the active center in the σ-form, in which the Nd atom is linked to the C_γ atom. This results in an increased content of *trans-1,2-* and *trans-1,4* units in polypiperylene.

In the polymerization of isoprene, high regioselectivity and the absence of *trans-1,4*-units in the polymer are related to the hampered rotation of the $C(CH_3){=}CH_2$ double bond of isoprene incorporated into the complex with the active center.

THE ROLE OF ANTI-SYN ISOMERIZATION OF THE TERMINAL UNIT OF THE GROWING POLYMER CHAIN

The possibility for the occurrence of *anti-syn* isomerization was qualitatively estimated from the experimental dependence of the relative content of *cis-* and *trans*-units in the polymer on the monomer concentration in polymerization with *cis*-regulating systems.[4] Similar dependence was also observed for lanthanide catalysts.[84] To obtain more detailed information on the reaction of isomerization, we performed a quantitative estimation of the kinetic scheme of the propagation reaction, which includes *anti-syn-* isomerization, for polymerization butadiene and piperylene with the $NdCl_3 \cdot 3TBP\text{-}Al(i\text{-}C_4H_9)_3$ system.[72,77,86-89] According to the data of quantum-chemical studies, which indicates that the isomeric forms of active centers (in butadiene polymerization) must not markedly differ by their reactivity, the following scheme of the reaction was analyzed:

$$
\begin{array}{ccccc}
 & K_1 & & K & \\
\wedge\wedge\wedge AS + M & \longrightarrow & \wedge\wedge\wedge AS\cdot M & \longrightarrow & \wedge\wedge\wedge cis\text{-}1,4\text{-}unit\text{-}AS + \\
 & K_{-1} & & & \\
K_3 \downarrow\uparrow K_4 & & K_5\downarrow\uparrow K_6 & & \\
 & K_1 & & K & \\
\wedge\wedge\wedge AS_t + M & \longrightarrow & \wedge\wedge\wedge AS_t\cdot M & \longrightarrow & \wedge\wedge\wedge trans\text{-}1,4\text{-}unit\text{-}AS + \\
 & K_{-1} & & &
\end{array}
$$

where $\wedge\wedge\wedge$AS is the *anti*-structure of active center, AS·M – is a complex of active center with monomer. The analogous designations are used for the *syn*-structure (AS_t) of active center.,

The meaning of the constants (K_1, K_{-1}, K_3, K_4, K_5 and K_6) can be deduced from the scheme. In accordance with scheme it is possible to derive the following approximate expressions for chain propagation rate (V) and the ratio between *cis*-1,4- and *trans*-1,4-unit formation rates (V_{cis} and V_{trans}, respectively):

$$V = K_pC_\alpha[M] \tag{1}$$

$$V_{cis}/V_{trans} = \gamma + [M]/(\alpha_2[M] + \beta_2) \tag{2}$$

where $K_p = 1/(\alpha_1[M] + \beta_1)$;

$\gamma = K_6/K_5$; $\beta_1 = (K + K_{-1})/KK_1$; $\alpha_1 = 1/K$;

$$\alpha_2 = \alpha_1 K_5;\ \beta_2 = K_3[\ \alpha_2 (1 + \gamma) + \beta_1 K_1]/K_1 \qquad (3)$$

This scheme takes into account that propagation is a two-stage reaction (coordination and insertion) and that *anti-syn* isomerization of the terminal unit of the growing chain can occur both prior to diene coordination to the active center and after formation of the complex between diene and the active center.

The meaning of the constants is clear from the scheme. Formation of 1,2-units in polybutadiene was not taken into account because in all experiments, their content did not exceed 1-3%. The value of γ, α_2, and β_2 parameters were determined by minimizing the weighed sum of the squares of deviations of the experimental values of V_{cis}/V_{trans} from the theoretical values determined from expression (2). Table 5 list the experimental values of the constants for separate stages of chain propagation in butadiene polymerization with the $NdCl_3 \cdot 3TBP\text{-}Al(i\text{-}C_4H_9)_3$ system.

Table 5. The Value of Rate Constants for Elementary Stages of Propagation Reactions and their Ratios in Polymerization of Butadiene with the Catalytic Systems $NdCl_3 \cdot 3TBP\text{-}Al(i\text{-}C_4H_9)_3$ (I) and $NdCl_3 \cdot 3TBP\text{-}Al(i\text{-}C_4H_9)_2H$ (II).

Catalytic system	Solvent	$\beta_1 = KK_1/(K + K_{-1})$, l/(mol min)	$\alpha_1^{-1}=K$, min^{-1}	$\alpha_2 = K_6/K$	K_3, min^{-1}	K_6, min^{-1}	β_2^*, mol/l	γ
I	Heptan	2670	8000	0.08	22	660	0.01	0.98
	Toluene	900	-	0.04	80	-	0.09	0.97
II	Toluene	380	-	0.09	15	-	0.04	0.93

Note: $^*\beta_2 \approx K_3(2\alpha_2/K_1 + \beta_1)$.

As is seen, the nature of solvent most dramatically affects the stage of coordination (constant K_1 in toluene is ~3 times smaller than that in heptane) and the stage of isomerization (constant K_3 changes by ~ 4 times on going from toluene to heptane). This may be associated with the fact that in toluene, the active centers not complexed with the monomer form complexes (of the arene type) with aromatic hydrocarbon. The occurrence of aromatic hydrocarbon in the coordination sphere of Nd, first, hampers the coordination of the monomer (K_1 decreases) and, second, favors the activation of π-σ transitions (i.e., increases the rate of *anti-syn* isomerization, which involved the σ-form of active center). This is reflected in an increase in the magnitude of K_3. For the centers complexed with the monomer, the isomerization constant (K_6) further increases, indicating that the monomer activates π-σ transitions to a greater extent than the solvent. Thus, in heptane, the isomerization constant for the active centers complexed with the monomer (K_6) is 30 times larger than the isomerization constant for free active centers (K_3).

CONCLUSION

The suggested mechanism of *cis*-regulation takes into account in an explicit form the kinetic factor (time of occurence of the terminal unit in the π- and σ-structures) and readly permits the following general conclusion to be drawn. In the case where the time of occurence of the terminal unit in the π-structure exceeds time τ_1 (reaction time), the stabilization (fixation) of the σ-structure of active center by the diene molecule would be no longer neccessary and the diene molecule can participate in the reaction regardless of its conformation (*cis*- or *trans*-). Naturally, this should imply the loss of *cis*- stereospecificity of active center. On the other hand, if the system does not possess *cis*-regulating properties (i.e. the time of occurence of the terminal unit in the σ-structure exceeds τ_1), the stabilizing π-structures (for example, by introduce of halogenating agent) could lead to the emergence of *cis*-stereospecificity of active centers.

Thus, the major components of active centers (such as terminal unit, bridging bond, organoaluminum component, lanthanide atom and its ligands) can decisively affect its stereospecificity. The role of each element manifests itself through different factors, and the possible mechanisms of their influence were discussed above. The key point determining the value of one or another factor in the mechanism of stereoregulation is the relationship between the following characteristic times: time of occurence of the terminal unit in the π-state, time of occurence of the terminal unit in the σ-state, time of the elementary event of insertion, time of the elementary event of isomerization, time interval between two successive additions of diene molecules to polymer chain.

The above said shows how the interplay of these times can affect the microstructure of the resulting polydiene. Naturally, real active centers are markedly more complicated, and the mechanism of stereoregulation apparently includes many details, which are yet to be studied. However, account of the characteristic times for different stages of the complicated reaction of chain propagation will undoubtedly always be the necessary stage for both the study of all diverse possibilities of lanthanide catalysts and the development of new catalytic systems.

ACKNOWLEDGEMENT

We are grateful to the Russian Foundation for Basic Research.(grant no. 99-03-33437) for support of this research. We also thanks our coworkers Dr.N.Marina and Dr.V.Urazbaev for valuable discussions.

REFERENCES

[1] B.A.Dolgoplosk and E.I.Tinyakova, *Polymer Science*, V.**36A**, №10, p.1380-1403 (1994).
[2] L.Porri, A.Giarusso and G.Ricci, *Polymer Science*, V.**36A**, №10, p.1421-1432 (1994).
[3] Yu.B.Monakov, N.G.Marina and Z.M.Sabirov, *Polymer Science*, V.**36A**, №10, 1404-1420 (1994).

[4] B.A.Dolgoplosk and E.I.Tinyakova. *Metalloorganicheskii katalyz v protsessakh polimerizatsii (Organometallic catalysis in polymerization processes).* Moscow: Nauka, 2-nd ed, 536 p. (1985).
[5] Yu.I. Murinov, Yu.B.Monakov and Z.G. Shamaeva *et al., Izv.ANSSSR, Ser.Khim.*, №12, p.2790-2792 (1977).
[6] M.V.Eskina, V.A. Lovchikov and A.S. Khachaturov *et al., Vysokomol. Soedin.*, V.**28B**, №12, p.887-890 (1986).
[7] K.D.Skuratov, M.I. Lobach and A.S. Khachaturov *et al., Vysokomol. Soedin.*, V.**29A**, № 7, p.1406-1412 (1987).
[8] A.A. Sanyagin and V.A. Kormer, *Dokl.ANSSSR*, V.**283**, № 5, p.1209-1211(1985).
[9] N.G. Marina, N.V. Duvakina and Yu.B. Monakov *et al., Vysokomol. Soedin.*, V.**27A**, № 6, p.1203-1207 (1985).
[10] Yu.B. Monakov, N.G.Marina and G.A.Tolstikov, *Chemia Stosowana*, vol.**32**, №3/4, p.547-557 (1988).
[11] Yu.B. Monakov, N.G.Marina and R.M.Khairullina *et al., Inorg.Chim.Acta*, vol.**142**, № 1, p.161-164 (1988).
[12] V.G.Kozlov, N.G.Marina and I.G.Savel'eva *et al., Inorg.Chim.Acta*, vol.**154**, № 2, p.239-245(1988).
[13] V.I.Aksenov, V.B.Murachev and V.P.Ziborova *et al., Vysokomol. Soedin.*, V.**30B**, № 6, p.431-436(1988).
[14] Yu.B.Monakov, N.G.Marina and G.A.Tolstikov, *Polymery*, vol.**34**, № 6-7, p.263-270 (1989).
[15] R.M.Khairullina, N.G.Marina and O.I.Kozlova *et al., Vysokomol. Soedin.*, V.**33B**, № 6, p.463-465 (1991).
[16] Yu.B.Monakov, N.G.Marina and O.I.Kozlova *et al., Dokl. ANSSSR*, V.**292**,№ 2, p.405-407 (1987).
[17] M.I.Lobach, I.A.Poletaeva and A.S.Khachaturov *et al., Polymer*, vol.**18**, № 11, p.1196-1198 (1977).
[18] S.V.Bubnova, A.I.Tverdov and V.A.Vasil'ev, *Vysokomol. Soedin.*, V.**30A**, № 7, p.1374-1379 (1988).
[19] S.R.Rafikov, Yu.B.Monakov and Ya.Kh.Bieshev *et al., Dokl. ANSSSR*, V.**229**, № 5, p.1174-1176 (1976).
[20] Yu.B.Monakov, Ya.Kh.Bieshev and A.A.Berg *et al., Dokl. ANSSSR*, V.**234**, № 5, p.1125-1127 (1977).
[21] Yu.B. Monakov, N.G.Marina and N.V.Duvakina *et al., Dokl. ANSSSR*, V.**236**, № 3, p.617-619 (1977).
[22] A.A.Panasenko, V.N.Odinokov and Yu.B.Monakov *et al., Vysokomol. Soedin.*, V.**19B**, № 9, p.656-658 (1977).
[23] A.A.Berg, Yu.B.Monakov and V.P.Budtov *et al., Vysokomol. Soedin.*, V.**20B**, № 4, p.295-299 (1978).
[24] I.N.Markevich, O.K.Sharaev and E.I.Tinyakova *et al., Dokl. ANSSSR*, V.**268**, № 4, p.892-896 (1983).
[25] V.A.Yakovlev, E.L.Vollershtein and L.S.Cherezova *et al., Dokl. ANSSSR*, V.**268**, № 6, p.1422-1425 (1983).

[26] O.G.Avdeeva, I.N.Markevich and O.K.Sharaev *et al., Dokl. ANSSSR*, V.**286**, № 3, p.641-643 (1986).
[27] G.A.Gailyunas, R.Kh.Biktimirov and R.M.Khairullina *et al., Dokl. ANSSSR*, V.**295**, № 6, p.1385-1388 (1987).
[28] O.G.Avdeeva, O.K.Sharaev and I.N.Markevich *et al., Vysokomol. Soedin.*, V.**32A**, № 2, p.367-373 (1990).
[29] B.A.Dolgoplosk, E.I.Tinyakova and I.N.Markevich *et al., J.Organomet. Chem.*, vol.**255**, № 1, p.71-79 (1983).
[30] E.L.Vollershtein, N.N.Glebova and S.B.Gol'shtein *et al., Dokl. ANSSSR*, V.**284**, № 1, p.140-142 (1985).
[31] I.Sh.Guzman, N.N.Chigir and O.K.Sharaev *et al., Dokl. Akad. Nauk SSSR*, V.**249**, № 4, p.860-862 (1979).
[32] B.A.Dolgoplosk, E.I.Tinyakova and I.Sh.Guzman *et al, J.Organomet. Chem.*, vol.**201**, № 1, p.249-255 (1980).
[33] N.N.Chigir, I.Sh.Guzman and O.K.Sharaev *et al., Dokl. ANSSSR*, V.**263**, № 2, p.375-378 (1982).
[34] E.L.Vollershtein, V.A.Yakovlev and E.I.Tinyakova *et al., Dokl. ANSSSR*, V.**250**, № 2, p.365-366 (1980).
[35] A.B.Nikitin, V.A.Yakovlev and B.A.Dolgoplosk, *Dokl. ANSSSR*, V.**291**, № 2, p.393-395 (1986).
[36] E.N.Zavadovskaya, O.K.Sharaev and G.K.Borisov *et al., Dokl. ANSSSR*, V.**274**, № 2, p.333-334 (1984).
[37] E.N.Zavadovskaya, O.K.Sharaev and G.K.Borisov *et al., Dokl. ANSSSR*, V.**284**, № 1, p.143-144 (1985).
[38] S.B.Gol'shtein, V.A.Yakovlev and G.N.Bondarenko *et al., Dokl. ANSSSR*, V.**289**, № 3, p.657-660 (1986).
[39] S.B.Gol'shtein, V.A.Yakovlev and A.I.Mikaya *et al., Dokl. ANSSSR*, vol.**309**, № 5, p.1129-1131 (1989).
[40] B.A.Dolgoplosk, V.A.Yakovlev and E.I.Tinyakova *et al., Dokl. AN (Russia),* V.**328**, № 1, p.58-60 (1993).
[41] Yu.B.Monakov, N.G.Marina and I.G.Savel'eva *et al., Dokl.ANSSSR*, V.**265**, № 6, p.1431-1433 (1982).
[42] S.R.Rafikov, V.G.Kozlov and N.G.Marina *et al., Izv.AN SSSR, Ser.Khim.*, 1982, № 4, p.871-875 (1982).
[43] V.A.Kormer, V.A.Vasil'ev and S.V.Bubnova *et al., Kauchuk i Rezina*, № 1, p. 5-8 (1986).
[44] Z.Shen, J.Ouyang and F.Wang *et al., J.Polym. Sci., Part A: Polym.Chem.*, vol.**18,** № 12, p.3345-3357 (1980).
[45] G.Lugli, A.Mazzei and S.Poggio, *Makromol.Chem.*, vol.**175**, № 7, p.2021-2027 (1974).
[46] N.G.Marina, Kh.K.Gadeleva and Yu.B.Monakov *et al., Dokl. ANSSSR*, V.**284**, № 1, p.173-175 (1985).
[47] G.M.Chernenko, V.A.Yakovlev and E.I.Tinyakova *et al., Vysokomol. Soedin.*, V.**31B**, № 8, p.637-638 (1989).
[48] N.N.Chigir, O.K.Sharaev and E.I.Tinyakova *et al., Vysokomol. Soedin.*, V.**25B**, № 1, p.47-48 (1983).

[49] N.G.Marina, Kh.K.Gadeleva and Yu.B.Monakov *et al., Dokl. ANSSSR*, V.**274**, № 3, p.641-643 (1984).
[50] V.G.Kozlov, R.V.Nefed'ev and N.G.Marina *et al., Dokl. ANSSSR*, V.**299**, № 3, p.652-656 (1988).
[51] N.G.Marina, N.V.Duvakina and Z.M.Sabirov *et al., Polymer Science*, vol.**39B**, № 1-2, p.34-35 (1997).
[52] G.Lugli, W.Marconi and A.Mazzei *et al., Inorg.Chim.Acta*, vol.**3**, №2, p.253-254 (1969).
[53] K.H.Thiele, R.Opitz and E.Z.Koehler, Z. *Anorg.Allg.Chem.*, vol.**435**, №8.p.45-48 (1977).
[54] Yu.B.Monakov, N.G.Marina and I.G.Savel'eva *et al., Dokl. ANSSSR*, V.**278**, № 5, p.1182-1184 (1984).
[55] V.S.Bodrova, E.P.Piskareva and S.V.Bubnova *et al., Vysokomol. Soedin.*, V.**30A**, № 11, p.2301-2306 (1988).
[56] S.V.Bubnova, E.P.Piskareva and V.K.Vasil'ev *et al., Vysokomol. Soedin.*, V.**35B**, № 1, p.18-21 (1993).
[57] V.S.Bodrova, E.P.Piskareva and V.A.Kormer, *Dokl. ANSSSR*, V.**293**, № 3, p.645-649 (1987).
[58] H.L.Hsieh and G.H.C. Yeh, *Industr.Eng.Chem.Product.Developm.*, vol.**25**, № 3, p.456-463 (1986).
[59] X.Zhang, F.Pei and X.Li *et al., Acta Polym. Sinica*, vol.**8**, № 4, p.391-398 (1990).
[60] Y.Jin, X.Zhang and F.Pei *et al., Chinese J.Polym. Sci.*, vol.**8**, № 2, p.121-126(1990).
[61] J.Wei, Y.Liao and, *Gaofenzi Tongxun*, № 5, p.342-346(1983).
[62] Y.T.Jin, Y.F.Sun and Z.Liu *et al., Fenzi Kexue Yu Huaxue Yanjju*, vol.**4**, № 2, p.247-252(1984).
[63] X.Zhao and F.Tian, *Kexue Tongbao*, vol.**27**, № 12, p.731-732 (1982).
[64] X.Zhao and F.Wang, *Kexue Tongbao*, vol.**28**, № 6, p.776-778 (1983).
[65] K.D.Skuratov, M.I.Lobach and A.N.Shibaeva et al., *Polymer,* vol.**33**, № 24, p.5202-5212 (1992).
[66] Y.Jin, X.Li and Y.Sun *et al., Gaofenzi Tongxun*, № 5, p.358-362 (1984).
[67] Y.Jin, X.Li and Y.Sun *et al., Kexue Tongbao*, vol.**30**, № 8, p.1047-1049(1985).
[68] C.Shan, Y.Lin and J.Ouyang *et al., Makromol.Chem.*, vol.**188**, p.629-635(1987).
[69] M.C.Throckmorton, *Kautsch.Gummi, Kunstst.*, vol.**22**, № 6, p.293-297 (1969).
[70] Y.B.Monakov, N.G.Marina and Z.M.Sabirov *et al., Dokl. ANSSSR*, V.**327**, № 4-6, p.524-527(1992).
[71] Yu.B.Monakov, Z.M.Sabirov and N.Kh.Minchenkova, *Mendeleev Commun.*, № 3, p.85-87 (1993).
[72] Z.M.Sabirov, N.Kh.Minchenkova and Yu.B.Monakov, *Vysokomol.Soedin.*, V.**32B**, № 2, p.83-84(1991).
[73] Z.M.Sabirov, Yu.B.Monakov, *Dokl.ANSSSR*, V.**302**, № 1, p.143-145 (1988).
[74] Z.M.Sabirov, Yu.B.Monakov and G.A.Tolstikov, *J.Mol.Catal.*, vol.**56**, № 1-3,p.194-202 (1989).
[75] Z.M.Sabirov and Yu.B.Monakov, *Inorg.Chim.Acta*, vol.**169**, №2, p.221-224(1990).
[76] Z.M.Sabirov, N.Kh.Minchenkova and Yu.B.Monakov, *Vysokomol.Soedin.*, V.**31A**, №11, p.803-804 (1990).

[77] Z.M.Sabirov, N.Kh.Minchenkova and I.I.Kazantseva *et al., Teor.Eksp.Khim.*, V.**27**, №6, p.741-745 (1991).

[78] Z.M.Sabirov, V.N.Urazbaev and Yu.B.Monakov, Polymer Science, V.**34A**, №5, p.379 (1992).

[79] Yu.B.Monakov, Z.M.Sabirov and V.N.Urazbaev, *Dokl.AN (Russia)*, V.**332**, №1, p.50-53 (1993).

[80] Z.M.Sabirov, V.N.Urazbaev and Yu.B.Monakov, *Polymer Science*, vol.**39B**, №1-2, p.22-25 (1997).

[81] Z.M.Sabirov and A.A.Bagatur'yants, *Usp.Khim.*,V.**60**, № 10, p.2065-2088 (1991).

[82] N.G.Marina, Yu.B.Monakov and G.A.Tolstikov, *Chem.Rev.*, vol.**18,** № 4, p.1-45(1994).

[83] D.A.G.Compton, W.O.George and W.F.Maddams, J.*Chem.Soc., Perkin Trans.2*, p.1311-1320 (1977).

[84] Yu.B.Monakov and G.A.Tolstikov, *Kataliticheskaya polimerizatsiya 1,3-dienov (Catalytic polymerization of 1,3-dienes)*, Moscow: Nauka, 211 p. (1990).

[85] Z.M.Sabirov, N.Kh.Minchenkova and V.N.Urazbaev et al., *J.Polym.Sci., Polym.Chem.Ed.,* vol.**31**, p.2419-2421(1993).

[86] Z.M.Sabirov, N.Kh.Minchenkova and Yu.B.Monakov, *Inorg.Chim.Acta*, vol.**160**, №1, p.99-101 (1989).

[87] Z.M.Sabirov, N.Kh.Minchenkova and N.A.Vakhrusheva *et al., Dokl.ANSSSR*, V.**312**, №1, p.147-150 (1990).

[88] Yu.Monakov, Z.M.Sabirov and N.G.Marina, *Polymer Science*, vol.**38A**, №3, p.232-240 (1996).

[89] Z.M.Sabirov, V.N.Urazbaev and N.G.Marina *et al., Polymer Science*, vol.**39B**, №1-2, p.14-17 (1997).

Chapter 4

The Estimation of Biological Consequences of the Combined Effect of the Surfactant and the Acute Irradiation at Low Dose on Mice by Using of Parameters of Lipid Peroxidation Regulatory System

L. N. Shishkina, Ye. V. Kushnireva and M. A. Smotryaeva
N.M. Emanuel Institute of Biochemical Physics, Russian Academy of Sciences, ul. Kosygin, 4, Moscow, 119991, Russia

Abstract

It is found that intraperitoneal administration of the Tween 80 at the 300 mg/kg dose in mice effects on the absolute values of the lipid peroxidation (LPO) regulatory system parameters but the interrelation between these parameters keeps up just the same the control mice. The effect of surfactant on the intensity of LPO, the antioxidant status and lipid composition of liver and DNA content in blood plasma of mice was studied. The combined effect of Tween 80 at the 30 mg/kg dose and X-irradiation at the dose of 16 cGy (16 R) causes the disturbances in functioning of LPO regulatory system. The effect of these disturbances depends on the time after irradiation, the fact whether Tween 80 is administered before or after irradiation, what parameter is analyzed and what tissue is investigated. The administration of Tween 80 after irradiation intensified the effect of X-rays at low dose in more extent than its administration before irradiation.

Aims and Background

The impossibility to predict the combined effect of the different factor occurs due to the maximum probability of appearance of synergistic effect from different chemical and physical factors at low doses and intensities [1-5]. The research of the combined effect of the different environmental factors on organisms is a high priority problem due to the Chernobyl disaster, atomic bomb tests, radioactive waste concealment, chemical weapon tests and destruction. The surfactants are among the technogenic unfavourable factors. Their role for the development of the biological consequences of the combined effect of the different agents enhances due to the surfactant's properties to facilitate the penetration of hydrophobic compounds into cells. In practice, this distinction of surfactants common is used for the prophylactic or therapeutic administration of the lipid soluble substances into organisms. Usually, polyoxyethylensorbitanmonooleate (Tween 80) is used for the hydrophobic antitumor and radioprotective agent administration [6, 7]. It exerts the little cytotoxic effect at the minor doses [8], especially in mice and rats when given by intraperitoneally (i.p.) and per os [9]. Tween 80 also has no teratogenic effect in Wistar rats when were given at a dose of 0.1; 1.0 or 10 % in the diet [10]. However, in a number of articles both Tween 80 alone and Tween 80 in combination with the different agents reveal the biological activity on cell and organism levels [11, 12]. It is shown that it can nether be used as a solvent for experiments when considered for intravenous administration, and it can be employed as vehicle for neuropsychopharmacological experiments in doses not exceeding 1 ml/kg.

The aim of our research was to study of the Tween 80 influence on the intensity of LPO process in liver and blood plasma, antioxidative status and lipid composition of liver and DNA content in blood plasma when surfactant administrated before or after X-irradiation of animals at low dose.

EXPERIMENTAL

The experiments were carried out on 115 mice SHK (females, 20 – 32.5 g) and 30 mice SHK (males, 17 - 24 g). Mice (males) were divided in two groups. The first group (15 mice) was used as intact biological control, the Tween 80 at the dose of 300 mg/kg as 3% solution was administration intraperitoneally at the second group of mice. Decapitation of mice was performed within 1 h and 1 day after the surfactant administration. The single acute X-irradiation of mice (females) at the dose of 16 R was performed by a RUT-200-20-3 apparatus in special plastic containers per 10 mice at the following conditions: current strength - 15 mA, filter 0.5 mm Cu, the dose rate - 44 R/min, the total irradiation interval – 22 s.

Mice SHK (females) were divided in 4 groups as follows:

1. The acute irradiation of mice at the dose of 16 R (group 1, 20 mice).
2. The prophylactic administration of Tween 80 as 0.3% solution within 30-45 min before mice irradiation (group 2, 40 mice).
3. The therapeutic administration of Tween 80 as 0.3% solution within 10-15 min after irradiation (group 3, 40 mice).
4. Mice of group 4 were the intact biological control.

Decapitation of mice of group 1 was done within 1 day and 1 month after irradiation. Decapitation of mice of groups 2 and 3 was done within 1 month after irradiation. Decapitation of intact mice was performed simultaneously with the experimental group per 6 - 8 animals. All parameters of intact mice were determined for each animal. Mice in experimental groups were divided in subgroups per 1 - 4 animals. After decapitation of mice, blood was collected in test tubes treated by 5% solution of sodium citrate. The blood plasma was separated by centrifugation. The murine liver was placed in ice-cooled weighting bottles immediately after decapitation.

Lipids were isolated by the method of Blay and Dyer in the Kates modification [13]. The antioxidative activity (AOA) of lipids was determined by using the methyl oleate oxidative model [7] as the ratio of the difference in the induction period of methyl oleate oxidation in presence and absence of lipids to concentration of the added lipids. Preliminary methyl oleate was purified by vacuum distillation. The lipid concentration in methyl oleate was 20 mg/ml. A specimen dissolved in methyl oleate was placed in an oxidizing cell with a glass porous filter. Oxidation was carried out in a temperature-controlled chamber at a constant temperature 37° ± 0.1°C by blowing air through at a rate providing oxidation in a kinetic range. The course of oxidation was followed by iodometric determination of peroxides formed. As the duration of the oxidation induction period we took the time in which the concentration of peroxides in the system reached 0.02 mmol/g and as a standard we chose methyl oleate with initiation rate of 0.29x10 M.s and an oxidation induction period 20 h [14]. The extent of oxidation inhibition was evaluated as the ratio of the oxidation induction period of experimental sample to the oxidation induction period of pure methyl oleate to lipid concentration in methyl oleate [15].

The qualitative and quantitative composition of phospholipids (PL) was determined by thin-layer chromatography as it was described in Ref. 16. It was used type G or H silica gel (Sigma, USA), glass plates 9 x 12 cm and mixture solvents chloroform: methanol: glacial acetic acid: water (25:15:4:1) as a mobile phase. All solvents were of especially pure or chemically pure grade. The development of chromatograms was performed by iodine vapour. For the colour reaction for phosphorus we used ammonium molybdate and ascorbic acid (Serva, FRG) and also perchloric acid of chemically pure grade. The amount of inorganic phosphorus was judged the optical density of the solutions as the wavelength λ = 810 nm with Beckman Du-50 (USA) instrument. In addition to quantitative analysis of the different fractions of PL the following parameters of lipid composition were also evaluated: PL proportion in the total lipid content (% PL), the ratio of phosphatidyl choline to phosphatidyl ethanolamine content in PL (PC/PE) and the ratio of the sums of the more readily to the more poorly oxidizable PL (ΣROPL/ΣPOPL). The ΣROPL/ΣPOPL value was calculated by the formula: ΣROPL/ΣPOPL = (PI+PS+PE+CL+PA)/(LPC+SM+PC), where PI is phosphatidyl inositol, PS is phosphatidyl serine, CL is cardiolipin, PA is phosphatidic acid, LPC are lysoforms of PL, SM is sphingomielin.

The content of LPO products which have reacted with 2-thiobarbituric acid (TBA-RS) was determined spectrophotometrically with adding 10 μl of 0.01% 4-methyl-2,6-ditert, butylphenol (BHT) solution in ethanol by method [17]. Protein was determined according to [18]. The content of cell-free DNA concentration in blood plasma was determined by fluorescence technique [19, 20] with the help of Hitachi M 850 spectrofluorimeter. Blood plasma (1 ml) was mixed with 1 ml of 4',6-diamidino-2-phenylindole dihydrochloride (DAPI) solution (0.4 μg/ml). The mixture was incubated for 30 min at 37° C since the whole binding of DNA and DAPI was within 15 - 30 min according to Ref. 21. The fluorescence measurements were carried out on a spectrofluorimeter in a 0.02 M solution of disodium EDTA, pH 10. The wavelength of the exciting UV-radiation was 338 nm, the fluorescence nm, the fluorescence spectra were measured at a wavelength of 457 nm, slit was 5 nm, the scanning rate was 60 nm/min. The amount of DNA (μg/ml) in the samples was determined by calibration curve, which was done by the native DNA of murine liver.

The data were processed with a commonly used variational statistic method [22] and by KINS program given in Ref 23. The variability of parameters was evaluated as the ratio of mean square error of average mean for group expressed as a percentage.

RESULT AND DISCUSSION

According to [24 - 26] steady state of LPO in the norm is due to the physico-chemical system regulating oxidative reaction in lipids (Fig. 1). Significant parameters of this system are AOA and composition of lipids, their oxidizability and structural rearrangements of membrane components. Hepatocytes which also have the high sensitivity to technogenic contamination of environment [27, 28] play a substantial role at the maintenance of organism homeostasis in the norm and under the development of many pathologies.

The investigation of Tween 80 influence on the LPO intensity, AOA and lipids composition was performed on murine liver because of the qualitative and quantitative characteristics of tissue homogenates presented the reaction organ as the whole on action and provided integrated appraisal about their LPO system state. Tween 80 was administered at a dose 300 mg/kg which was lower 3.6-fold comparatively with the dose when it can be employed safely as a tool for neuropsychopharmacological experiments [9]. These experimental data are given in Table 1. It is seen that surfactant has different effects just as an extent and a trend of changes in the LPO regulatory system parameters in murine liver so as a variability of action. The Tween 80 administration causes 1.5 times increase of the lipid AOA value in murine liver for a day so that the lipid AOA levels are practically equal within 1 h and 1 day after administration. Nevertheless, the parameter variability is similar for the experimental and control groups.

The TBA-RS amount is almost half as much in murine liver within 1 h after surfactant administration compared with one for control. It is well known that LPO intensity is evaluated by the TBA-RS product content in complex biological system [29]. Hence, the reduction of the LPO product amount with increasing the lipid AOA points to the fact that the LPO intensity in murine liver decreases for a day after surfactant administration. Generalized parameters of lipid composition of liver change to a lesser extent compared with the parameters of antioxidant status. There is a rise of the PL proportion in total lipid content, the greatest within 1 day after surfactant administration. On the other hand, the more readily oxidizable PL proportion in murine liver decreases since the ratio of ΣROPL/ΣPOPL is 20% lower than the same in the experimental groups compared with the control within 1 day after administration. Thus, the oxidizability of the liver lipid decreases after surfactant administration at this duration. The Tween 80 administration causes 2 and 2.5 times rise of the lysoform relative content in PL composition of murine liver within 1 h and 1 day after surfactant administration, correspondingly. Increasing LPC amount is noted commonly by the action of unfavourable factors on organism. That is revealed by acute irradiation of experimental animals at sublethal and lethal doses [30, 31] and also in liver PL composition of wild rodents which were caught in the Chernobyl NPP accident zone within the next few years [28].

Table 1. Influence of the Intraperitoneal Administration of Tween 80 at a Dose of 300 mg/kg on the Antioxidant Status and the Generalized Parameters of Lipid Composition in Murine Liver

Parameter	Control (n = 15)*	Period after administration of Tween 80	
		1 h (n = 8)	1 day (n = 7)
Lipid AOA (h.ml/g)	4500 ± 670	6750 ± 1600	7100 ± 1300
[TBA-RS] (nmol/mg of protein)	0.20 ± 0.03	0.116 ± 0.002	0.085 ± 0.07
% PL	36.7 ± 0.55	39.7 ± 1.6	45.8 ± 2.6
PC/PE ratio	2.33 ± 0.078	2.17 ± 0.064	2.55 ± 0.047
ΣROPL/ΣPOPL ratio	0.52 ± 0.032	0.49 ± 0.012	0.42 ± 0.008

* Mice number

Thus the intraperitoneal administration of Tween 80 at the dose of 300 mg/kg acts on the LPO regulatory system parameters in murine liver already after the surfactant administration. However, it should be noted that the control group of mice had a high antioxidant status of liver. This fact could influence the extent and trend of changes of biological system parameters. The dependence of extent of LPO regulatory system parameter changes on their initial state for control animals was previously established for such biological active substances as monoanimoxidase inhibitor and methyl uracil [32, 33].

The modification of the membrane permeability is one of the most sensitive function on action to low damaging factors [34]. This can cause significant variations of the cell-free DNA content. Indeed, it was earlier found a direct relation between the TBA-RS amount and the cell-free content in blood plasma of mice to the drinking water to which was added daily chemical toxic agent at different doses [35].

The TBA-reactive substance amount and the cell-free DNA content in blood plasma of mice within 26 h after irradiation at the dose of 16 R are shown in Table 2. The analysis of these data revels that DNA content in blood plasma of experimental and control mice is equal. However, the variability of this parameter in the experimental group is 2.3 times greater than in the control group. The TBA-RS quality in the blood plasma of experimental mice tends to increase as compared with the magnitude for the control group but the variability of TBA-RS amount in the both groups is similar (15.1% and 18.5% for the control and experimental groups, correspondingly). However, a closer examination of the obtained data revealed that the LPO intensity in blood plasma in individual within the experimental group varied in different direction. While the TBA-RS amount rises about 1.6 times for 70% of mice from experimental group, there are 30% of mice from the same group which amount of TBA-RS decreases 3.8 times. It should be emphasized that the parameter heterogeneity, i.e. the unsteadiness of the magnitude and direction of effect under fixed experimental conditions, is a principal peculiarity of the response of complex biological system on low action [36].

Table 2. The TBA-Reactive Substances Quantity and the Cell-Free DNA Content in Blood Plasma of Mice within 26 h after Irradiation at the Dose of 16 R

Variant of experience	[TBA-RS] (nmol/mg of protein)	[DNA] (μg/mg of protein
Control (n = 8)*	0.096 ± 0.0145	0.057 ± 0.004
Experience (n = 10)	0.115 ± 0.021	0.056 ± 0.009

* Mice number.

At the duration average correlation are found between the LPO product amount and DNA content in blood plasma of mice as the correlation coefficients are 0.63 ± 0.21 and 0.45 ± 0.25 for control and experimental group, correspondingly. The comparison of these magnitudes shows that there is trend to the diminution of the correlation coefficient for the exposed mice. Apparently, this diminution is caused by higher stability of DNA content in blood plasma of mice nearly after their irradiation at low dose.

Investigations of antioxidant status and LPO intensity in liver and blood plasma and also the cell-free content in blood plasma of mice within 1 month after their irradiation at the dose of 16 R with combination of Tween 80 as the additional agent, which is administered both before and after irradiation, exhibit several interesting generalities. The lipid AOA values of murine liver for the experimental and control groups within 1 month after irradiation are presented in Table 3. It is shown that irradiation of mice at the dose of 16 R reliably decreased the lipid AOA of murine liver of exposed mice as compared with that of control group. Furthermore, lipids of exposed mice accelerated the methyl oleate oxidation. However, attention should be paid to the possibility that the prooxidative activity of lipid can be due to both the depletion of antioxidants in them and the high extent of lipid nonsaturation [37]. The lipid AOA of liver of mice to whom Tween 80 was administrated both before and after irradiation changed in opposite directions as compared with that for the control mice. On the one hand, the lipid AOA decreased significantly in a half of the experimental mice approximately and on the other hand this parameter increased also in a half of the experimental mice. Similar heterogeneity of lipid AOA of tissues was earlier observed in wild rodents which were caught in the Chernobyl NPP accident zone [28].

Table 3. The Values of Lipids AOA of Mice SHK (females) within 1 Month after Irradiation at the Dose of 16 R

Variant of experience	Lipid AOA (h.ml/g)
Intact biological control	990 ± 330 (n = 9)*
X- irradiation (group 1)	– 340 ± 170 (n = 9)
Tween 80 administration as 0.3% solution within 30 - 45 min before irradiation (group 2)	– 50 ± 230 (n = 20) 2390 ± 330 (n = 18)
Tween 80 administration as 0.3% solution within 10 - 15 min after irradiation (group 3)	– 100 ± 90 (n = 20) 2440 ± 350 (n = 19)

* Mice number

The extent of changes of LPO product amount in liver of the mice which were administered with a surfactant after irradiation are more substantial than in groups of mice exposed after surfactant administration. The analysis of correlative interrelations between the lipid AOA values and LPO product amounts in liver showed the occurrence of a reverse correlation between these parameters for the control groups of mice (R = – 0.87 ± 0.09) (Fig. 2) and absence of similar correlation for mice of the all experimental groups. It is interesting to note the existence of the direct correlation with the correlation coefficient R = 0.99 ± 0.005 between the lipid AOA scale of change and the LPO product relative content in liver of mice of groups 1 and 3 (only irradiation and administration of Tween 80 after irradiation) (Fig. 3). The generality has not been found for mice groups when surfactant was administered before their irradiation at low dose.

Data about the limits of variations of generalized parameters of PL composition in liver of intact and experimental mice within 1 month after irradiation are given in Table 4. It is seen that both irradiation at low dose and its combined action with surfactant affect these parameters. The variations of the PC/PE ratio tend to higher values especially by the therapeutic administration of surfactant while the limits of variations of ΣROPL/ΣPOPL ratio are much less for the experimental groups than for the control especially by the prophylactic administration of Tween 80. The reverse correlative dependence between the ratio of the sums of the more readily to more poorly oxidizable PL and the ratio of PC/PE in liver PL of control and experimental groups of mice were revealed (Fig. 4). The correlation coefficients had the followed values: R = – 0.95 ± 0.03 for the control group of mice, R = – 0.82 ± 0.09 for the groups 1 and 3 and R = – 0.92 ± 0.05 for the group 2. It is seen that the values of the linear regression coefficient are different for the control and experimental groups of mice. The more substantial differences of this coefficient value were found in liver PL of mice that were exposed to low dose or after the surfactant therapeutic administration. Obviously, both the irradiation at low dose and the Tween 80 administration after X-irradiation causes the more significant changes of the structural state of liver lipids than their oxidizability.

Table 4. Limits of Variations of the Generalized Parameters of PL Composition in Murine Liver within 1 Month after Irradiation of Animals

Variant of experience	Parameter of PL composition		
	% PL	PC/PE ratio	ΣROPL/ΣPOPL
Intact biological control	10.7 - 60.8	1.19 - 2.28	0.50 - 1.06
Irradiation at the dose of 16 R (group 1)	21.3 - 83.6	2.01 - 2.39	0.43 - 0.53
Tween 80 administration as 0.3% solution before irradiation (group 2)	26.5 - 77.8	1.64 - 2.38	0.45 - 0.66
Tween 80 administration as 0.3% solution after irradiation (group 3)	28.3 - 65.9	1.50 - 3.07	0.39 - 0.70

The data obtained for the LPO product content and the cell-free DNA content in blood plasma of mice within 1 month after irradiation are given in Table 5 as average values for each variant of experience. It is seen that the both the TBA-reactive substance quantity and DNA content in blood plasma of mice show only a tendency to increasing by irradiation because the variabilities of the both parameters are 3.2 times more for the experimental group than for the control group. However, there is a need to note that the variability level of the LPO product amount is 30% more than that of the DNA content in blood plasma of both intact and exposed mice.

Table 5. Influence of Tween 80 on the LPO Product Amount and the Cell-Free DNA Content in Mice Blood Plasma within 1 Month after Irradiation of Animals at the Dose of 16 R

Variant of experience	[TBA-RS] (nmol/mg of protein)	[DNA] (μg/mg of protein)
Intact biological control (n*= 6)	0.1915 ± 0.028	0.072 ± 0.008
Irradiation at the dose of 16 R (group 1, n = 3)	0.42 ± 0.20	0.117 ± 0.043
Tween 80 administration as 0.3% solution before irradiation (group 2, n = 12)	0.197 ± 0.037	0.050 ± 0.015
Tween 80 administration as 0.3% solution after irradiation (group 3, n = 12)	0.178 ± 0.011	0.55 ± 0.006

* Number of replicate measurements.

The LPO product quantities as the average in groups in blood plasma of mice to whom Tween 80 was administered before or after irradiation do not distinguish from that for the control within 1 month after their application. At the same time the DNA content in blood plasma of mice in groups 2 and 3 decreases but the variability of this parameter is approximately the same by the surfactant therapeutic administration as a control group. The variability of the DNA content is similar in blood plasma of mice that were exposed to low dose or after surfactant prophylactic administration.

Straight correlation have been established between the TBA-reactive substance amount and cell-free DNA content in the blood plasma of the control mice (R = 0.97 ± 0.02), of the exposed mice (R = 0.99 ± 0.01) and at the surfactant prophylactic administration (R = 0.93 ± 0.04) within 1 month after irradiation at low dose. The corresponding data are presented in Fig. 5.However, a similar correlation for the group of mice which were administered to Tween 80 after irradiation is absent. It is the same group of mice which has the lowest variability of both the TBA-RS quantity and cell-free DNA content at this time. Furthermore, there is a need to note that the linear regression coefficient value of this dependence for mice of the control and the experimental groups is different. Thus, irradiation leads to a decrease in the linear regression coefficient value for the exposed mice, while the surfactant prophylactic administration results in an increase of this parameter compared with that for the control group and particularly for the exposed mice.

The data obtained indicate the absence of the uniform response and the high sensitivity of LPO regulatory system parameters in murine liver and blood plasma to the combined action of surfactant and irradiation at low dose. However, the sensitivity of different parameters of this regulatory system is different. The interrelation between the lipid AOA value and their extent of oxidation inhibition of methyl oleate is the most resistant. This conclusion follows from the correlation analysis between the lipid AOA and their extent of oxidation inhibition in mice liver within 1 month after irradiation. It is found the coordinates of the corresponding lines are equal for the control and for all experimental groups under condition that the initiation rates of oxidation of the model substrate are also equal. The correlation coefficients are equal to one ($R = 1$) in all cases. The i.p. administration of Tween 80 at a dose of 300 mg/kg affects the absolute values of the LPO regulatory system parameters but the interrelation between these parameters remains the same like in control. The combined action of Tween 80 at a dose of 30 mg/kg and X-irradiation at low dose evokes the disturbance at the LPO regulation system functioning which depends significantly on the time after action, the surfactant administration before or after irradiation and the choice of the tissue to be studied. For example, Tween 80 administration both before and after irradiation has a similar effect with respect to the extent and the direction on the lipid AOA changes in liver of the exposed mice and also evokes an absence of correlation between the lipid AOA values and the LPO product amounts. However, most probably the extent of disturbance of interrelation between these parameters is different because a straight dependence between the scale of changes of lipid AOA values and the LPO product quantities in liver within 1 month was found only for exposed group of mice to which Tween 80 was injected after irradiation. The marked changes of cell-free DNA content in the blood plasma on mice are revealed for the later period after low damaging factor action. It is the therapeutic administration of Tween 80 that caused the compete disruption of the relation between the LPO product amount and the cell-free content in the blood plasma of mice for a long time after irradiation. Perhaps, surfactant, if it is injected after acute irradiation at low dose, intensifies the biological effect of radiation more than if it is injected before.

The data obtained and data described in [27, 28] make it possible to suppose that the high sensitivity of physico-chemical LPO regulatory system parameters to the action of environmental factors may be used as complex tests to estimate the consequences of unfavourable factor action on complex biological organisms.

ACKNOWLEDGYMENT

This work was supported by the International Science and Technology Center Program, Project No 547-98.

REFERENCES

[1] M.A. Harwell, T.C. Hutchinson et al.: *Environmental Consequences of Nuclear War. V.II. Ecological and Agricultural Effects*. (Russian version), Mir, Moscow, 1988.

[2] V.G. Petin, V.P. Komarov: Quantitative Description of the Modification Radiosensibility. *Energoatonizdat*, Moscow, 1988 (in Russian).
[3] A.F. Malenchenko, S.N. Sushko, S.I. Plotnikova et al.: *Dokl. Akad. Nauk BSSR*, **33**, 1043 (1989).
[4] A.F. Malenchenko, S.N. Sushko, T.S. Kuz'mina: *Radiat. biologiya. Radioekologiya*, **35**, 777 (1995) (in Russian).
[5] V.G. Petin, G.P. Zhurakovskaya, A.G. Pantyukhina, A.V. Rassokhina: *Radiat. biologiya. Radioekologiya*, **39**, 113 (1999) (in Russian).
[6] O.S. Frankfurt: *Vopr. onkol.*, **21**, 63 (1975) (in Russian).
[7] E.B. Burlakova, A.V. Alesenko, E.M. Molochkina et al.: *Bioantioxidants in Radiation Damage and Tumor Growth*. Nauka, Moscow, 1975 (in Russian).
[8] I. Tsujino, T. Yamazaki, M. Masutani et al.: *Cancer Chemother. Pharmacol.*, **43**, 29 (1999).
[9] R.K. Varma, R. Kausha, A.Y. Junnarkar et al.: *Arzneimittelforschung*, **35**, 804 (1985).
[10] V. Eva, T. Itami, H. Kawaski, S. Kanoh: *Drag Chem. Toxicol.*, **11**, 249 (1988).
[11] C.M. Brubaker, D.H. Taylor, R.J. Bull: *Life Sci.*, **30**, 1965 (1982).
[12] B. Dib, M. Falchi: *Int. Tissue React.*, **18**, 27 (1996).
[13] M. Kates: *The Technology of Lipidology* (Russian version). Mir, Moscow, 1975.
[14] L.N. Shishkina. In: *Investigation of Synthetic and Natural Antioxidants in vitro and in vivo*. Nauka, Moscow, p. 26, 1992 (in Russian).
[15] V.A. Men'shov, L.N. Shishkina, Z.N. Kishkovsky: *Appl. Biochem. Microbiol.*, **30** (3), 359 (1994).
[16] *Biological Membranes. A Practical Approach* (Eds. J.B.C. Findlay, W.H. Evans), Mir, Moscow, 1990 (Russian version).
[17] T. Asakawa, S. Matsushita: *Lipids*, **15** (3), 137 (1980).
[18] R. Itzhaki, D.M. Gill: *Anal. Biochem.*, **9**, 401 (1964).
[19] J. Kapuscinsci, B. Skoczylas: *Anal. Biochem.*, **83**, 252 (1977).
[20] J. Kapuscinsci, B. Scoczylas: *Nucleic Acids Research*, **5**, 3775 (1978).
[21] N.V. Blyukhterova, M.A. Smotryaeva, K.E. Kruglyakova: Izv. *Acad. Nauk Russia. Ser. biol.*, 3, 276 (1996).
[22] G.F. Lakin: *Biometry. Vys. shkola*, Moscow, 1990 (in Russian).
[23] E.F. Brin, S.O. Travin: *Chem. Phys. Reports*, **10**, 830 (1991).
[24] S.A. Aristarkhova, G.V. Arkhipova, E.B. Burlakova et al.: *Dokl. Akad. Nauk SSSR*, **228**, 215 (1976).
[25] E.B. Burlakova, G.V. Archipova, M.I. Djalyabova et al. In: *Biophysical and Biochemical Information Transfer in Aging and Recognition*. Plenum Press, New York, 1979, p. 583.
[26] Ye. B. Burlakova, N.P. Pal'mina, Ye.L. Mal'tseva. In: *Membrane Lipid Oxidation* (Ed. Carmen Vigo-Pelfrey). CRC Press, Boston, 1991, V. III. p. 209.
[27] L.N. Shishkina, L.D. Materii, A.G. Kudyasheva et al.: *Radiobiologiya*, **32**, 19 (1992).
[28] A.G. Kudyasheva, L.N. Shishkina, N.G. Zagorskaya, A.I. Taskaev: *Biochemical Mechanism of Radiation Damage of Wild Rodent Natural Population*. Nauka, S. Peterburg, 1997 (in Russian).

[29] V.E. Kagan, O.N. Orlov, L.L. Prilipko: Problems of Analysis of Lipid Peroxidation Endogenous Products. In: *Itogi nauki i tekhniki VINITI Akad.* Nauk SSSR. Ser. Biofizika, Moscow, 1986 (in Russian).
[30] E.B. Burlakova, G.V. Archipova, L.N. Shishkina et. al.: *Studia Biophys.* **53**, 65 (1975).
[31] A.I. Medvedev, Yu.V. Evtodienko, L.S. Yaguzhinsky, A.M. Kuzin: *Inform. Bull. Radiobiologiya*, **18**, 95 (1975) (in Russian).
[32] Ch. Kairane, E.A. Baumanis, E.B. Burlakova et al.: *Vopr. Med. Khemii*, **25**, 776 (1979) (In Russian).
[33] Yu. P. Taran, L.N. Shishkina: *Vopr. Med. Khemii*, **39**, 37 (1993) (un Russian).
[34] G.Y. Koteles: *Atom. Energy Rev.*, **17**, 3 (1979).
[35] E.V. Shtamm, L.N. Shishkina, N.B. Kozlova et al.: *Vodosnab. and sanitar. tekhnika*, 10, 18 (1997) (in Russian).
[36] E.B. Burlakova: *Herald RAS*, **64** (3), 205 (1994),
[37] L.N. Shishkina, V.A. Men'shov, E.F. Brin: *Izv. Akad. Nauk Russia. Ser. biol.*, 3, 292 (1996).

Chapter 5

BASALTIC CONTINUOUS MULTIFILAMENT YARNS AGING IN AGGRESSIVE ALKALINE MEDIA. KINETIC MODEL

A. A. Dalinkevich*, K. Z. Gumargalieva**, A. V. Sukhanov* and A. V. Aseev***

*Central Research Institute of Special Machine-Building, 1 Zavodskaya Street, Khotkovo, Moscow region, Russia,

**Inslitute of Chemical Physics, RAS, 4 Kosygin Street., Moscow 117334, GSP-1. Russia,

***Company "Armoproject", 27 Malakhitovaya street, Moscow 128129.

ABSTRACT

The kinetic regularities of the breaking force reduction and the mass loss at the basalt roving aging in the alkaline media have been investigated. The obtained equations take into account the reduction of the basaltic fiber breaking force caused by the occurrence of the adsorption processes and the partial chemical dissolution (etching) of the surface layer basalt components in the aqueous alkali solution. The obtained regularities can be used for the prognosis of the tensile strength reduction of basaltic fiber materials under prolonged action of the alkaline aggressive media.

INTRODUCTION

The basaltic fiber materials are perspective fillers for composites both with organic [1, 2] and with inorganic [3] matrixes. However, contrary to the widely used for polymer reinforcement glass fibers [4-6], the chemical stability of the basaltic fibers has been investigated to a lesser extent [1-3, 7]. Some of the experimental data on the basaltic fibers chemical stability and aging in the aggressive media are published in works [3, 7]. However, in the works, there is no kinetic analysis of both the fiber physical property change and the regularities of the physico-chemical processes occurring at interaction of the basaltic fiber surface with the aggressive alkaline media.

Because of this, the goal of this work was estimation of the kinetic regularities of the commercial continuous basaltic roving physical property change at aging in the alkaline media.

EXPERIMENTAL PART

In this work, the continuous basaltic roving made at "*Sudogodskoye Sleklovolokno,*" ("*Fiber glass of Sudogda*") was used. Its linear density was 1250 tex (i.e, the portion of roving length 1000 m has a mass equal to 1250 grams). The roving consists of a bundle of lightly-twisted multi-filament continuous yams. The average diameter of the single filament (monofilament) was ≈ 9 μm. In the process of roving preparation, the filaments surface was treated by the oiling agent of the direct action, i.e., it contained an oil hydrophobic component and the dressing organosilicon component (aminosilane adhesion promoter). The mass content of the oiling agent was ≈1 %. The chemical composition of the basaltic fibers as the percentage of the basalt oxides is given in Table 1.

Table 1. Chemical Composition of the Basaltic Filaments.

Oxide type	**SiO_2**	**Al_2O_3,**	**Fe_2O_3**	**CaO**	**MgO**	**TiO_2**	**Na_20+ K_20**	**MnO**	**Other contaminants**
Content, mass %	54.3	15.9	9.9	7.75	4.50	1.1	4.0	0.24	Till 100%

The chemical aging of the roving samples was conducted by keeping them in the NaOH solution concentration 1.25N for different times. For this purpose, the roving parts length ≈25 cm were placed in desiccators, filled them with the NaOH solution, and closed them. The exposition time was from 2 to 72 days. After each eight days, the alkali solution over the samples was replaced with the fresh one. After certain time, samples were taken out from the alkali solution and washed in the large amount of the running tap water. After that, the samples were washed in distilled water and dried out on a filter paper at room temperature.

Before the mechanical testing of the original and the aged samples, the sample ends were placed and glued between the strong paper layers (the sample ends were put into the tensile machine grips). The breaking force was measured by using the Instron 1122 machine at the grip movement 50 mm/min. The sample span was 180 mm. The breaking force for the original roving sample was $P_0 = 650 \pm 16$ N.

The samples for the gravimetric investigations after careful washing in the top and distilled water were dried out in an oven at 150°C for two hours. The mass change of the aged samples was determined on the analytic microbalance "Sartorius" with a preciseness 10^{-5} grams. The raster electronic microscope JSM-5300LB made by JEOL firm (Japan) was used for studying the effect of the alkaline media on the basaltic fiber materials.

RESULTS AND DISCUSSION

From the experimental data shown in Fig. la, it is seen that the kinetic curve of the mass loss $\Delta m = (m_0 - m)/m_0$ (where: m_0 and m are the initial and the current value of the basaltic roving mass, correspondingly) can be considered as consisting from three regions. The fast mass loss occurring in the region 1 and does not exceeding ≈1.1 mass % is connected basically with the removal of the oiling agent from the filament surface in the alkaline solution. The next, slowly occurring mass loss (region 2), is connected with the slow diffusion-adsorption penetration of the aggressive media into the microscopic and submicroscopic defects (pores and cracks) of the fiber surface layer, the realization of the chemical solution (etching) of basalt on the walls of the defects. At the end of the region 2, after the completion of the diffusion-adsorption processes of filling the microscopic defects in the surface layer and some alkaline etching of them, on the fiber surface, the modified layer of certain thickness is formed on the fiber surface. The layer has developed defects, i.e it has increased sizes of micropores and cracks. In the future (region 3), through this layer, it is occurred a practically stationary diffusion transport of water and the anions OH^- to the deeper fiber inner layers the reversed transport of the dissolved basalt components. As it is seen from Fig. la, in the region 3, the fiber etching rate is increasing and remains almost constant during the whole experiment time. The increase of the etching rate (the mass loss) of the basaltic fibers in the region 3 (Fig. la) can be connected (on the analogy with the glass fibers) with the development of the surface defects, i.e. with increasing the depths and size of the pores and microcracks diameter. This, in turn, will allow water and the anions OH^- to go deeper inside the fiber what will increase the basaltic fiber etching rate as it was found experimentally (region 3, Fig. la).

The comparison of the mechanical properties of the aged basaltic roving samples (Fig.lb) with the discussed data on their etching (mass loss) shows that the fastest and a significant decrease of the breaking force occurs in the regions 1 and 2. As it was discussed above, just in the regions, the adsorption-desorption processes - the oiling agent removal, diffusion-adsorption filling of the fiber surface defects by the aggressive media, and a partial etching of fibers - take place. Hence, the fast decrease of the fiber breaking force at the beginning of the aging has an adsorption nature ("Rebinder effect"), the same as in case of the glass fiber chemical aging [4, 8, 9]. It is obvious that the mass loss caused by dissolution of the basalt itself and the corresponding decrease of the fiber cross-section at this stage of aging (the regions 1 and 2, Fig. la) is very small and cannot be responsible for the observed significant decrease of the breaking force.

To describe the kinetics of the breaking force decrease (and correspondingly, the tensile strength because the area cross-section of the elementary fiber (i.e. monofilament) roving is practically constant on this stage of aging), it is possible to try to use the kinetic regularities known for the adsorption on the solids surface [10]. For the simplicity, let us assume that the used oiling agent does not create an effective protective film on the fiber (monofilament) surface, easily washing off, does not interfere the adsorption of the aggressive media, and, therefore, has a little effect on the kinetics of the adsorption decrease of the breaking force. Indeed, as it is seen from the data in Fig.l b, the made assumption does not contradict to the experiment. Let us accept that the kinetics of the adsorption filling of the surface layer of fibers by the aggressive medium can be described by the known [10] equation:

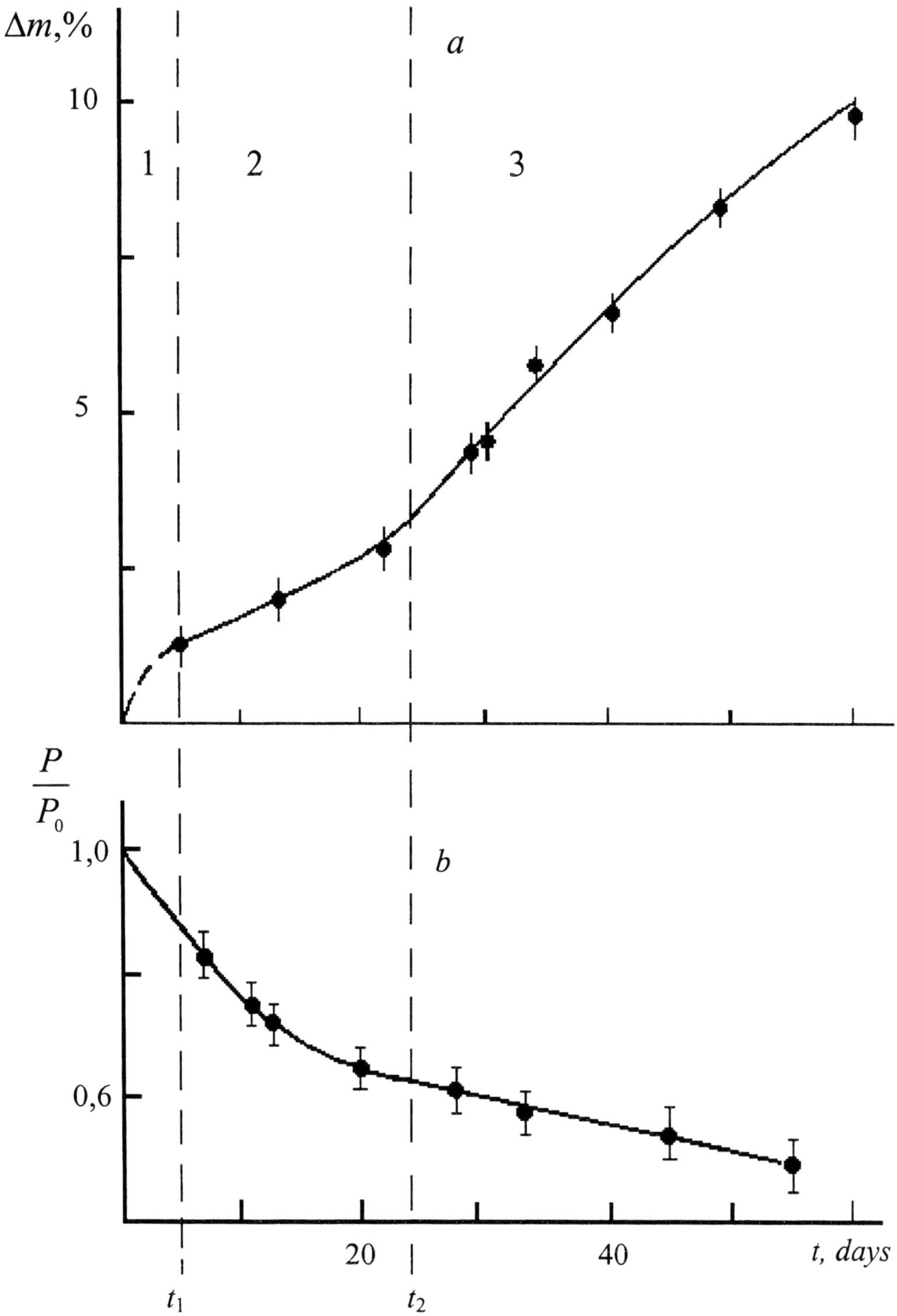

Figure 1. Kinetic curves of the mass loss (a) and the reduction of the breaking force (b) of the basaltic roving at aging in the aqueous 1.25 N solution of NaOH. The dashed lines are the borders of the regions 1, 2 and 3.

$$S=S_0 \exp(-k_s t) \qquad (1)$$

where: S_0 is the initial (at $t=0$) area surface of the fiber, which is free of the adsorbed aggressive medium; S is the current value of the free fiber surface; k_s is the constant of the surface interaction of basalt with the alkali aqueous solution (chemosorption); t is the exposure time of samples in the alkali solution.

Let us also accept that at this stage of the fiber breaking force reduction due to adsorption presses occurrence will go from P_0 (at $t=0$, $S=S_0$) to certain P_v corresponding to the fiber with a fully filled fiber surface. This state is achieved at large enough t, when the exponent in Eqn. (1) is approaching zero, and, thus, the P_v is corresponding to the volume component of the strength force. If we will consider that the decrease of the breaking force will occur from P_0 till a certain current value of P in the interval $P_v \le P \le P_0$ proportionally to the value of the absorptionally filled fiber surface, S_{ads}, then

$$P_o - P = \alpha \cdot S_{ads} = \alpha \cdot (S_o - S) = \alpha \cdot S_o (1 - \exp(-k_s \cdot t)) \qquad (2)$$

From the expression (2), by incorporation $t \to \infty$, we obtain:

$$\alpha S_o = P_o - P_v \qquad (3)$$

Finally, for the region of the fast adsorption decrease of the breaking force, we obtain:

$$\frac{P}{P_o} = \frac{P_o - P_V}{P_o} \cdot \exp(-k_s \cdot t) + \frac{P_V}{P_o} \qquad (4)$$

From the expression (4), it is seen that the breaking force can be represented as a sum of two components: the decreasing surface component and the volume component P_v / P_0. While deriving the equation (4), we assumed for the simplicity that the volume component P_v/P_0 = const., i.e. we neglected the slow dissolution of the fiber in the regions 1 and 2 (Figs1a and b). By taking into account the simultaneously occurring processes of adsorption and chemical dissolution of the fiber, the decrease of the breaking force for the whole experiment time will be:

$$\frac{P}{P_o} = \frac{P_o - P_v(0)}{P_o} \cdot \exp(-k_s \cdot t) + \frac{P_v(t)}{P_o} \qquad (5)$$

In the formula (5), the $P_v(0)$ is the value of the volume component at $t=0$. The type of the $P_v(t)$ dependence is determined by the kinetics of the diffusion-controlled alkali etching of the basaltic fiber.

The electron microscope investigations of the aged fibers showed that they consist of two weakly-connected between themselves layers. The outer, aged ring layer, consisting of the insoluble basalt components, has a loose, highly porous structure, and is easily penetrated by an electrolyte. The fiber inner cylindrical trunk is practically untouched by corrosion. At long enough aging time, between the layers, it was fond a clearly visible boundary and the separation of the outer layer fragments. With the aging time increase, the thickness of the aged outer ring layer was increasing, but the diameter of the inner cylindrical trunk, untouched by the aging and carrying the strength force of the aged sample, was decreasing. In general, the fiber summary diameter change with aging (up to 55 days) was in the experiment error range characteristic for the original fiber.

The low permeability of the fiber basalt block toward the aggressive media (water) and the relatively high rate of the catalyzed by alkali hydrolysis of some of the basalt component parts (the Si-O-Si, Al-0-Si, and partly Al-O-Al bonds [3]) allow to assume that the alkaline etching of the basalt fiber in the region 3 (Figs la and 1b) occurs in the outer diffusion-kinetic area [11], i.e. from the surface. Such regularities of the chemical aging of some glass fibers were noted in the monograph [8]. It should be also noted that the discussed above the electron-microscopic picture of the basaltic fiber morphological changes as the result of the alkaline aging is in agreement with the assumption about the fiber etching occurring in the outer diffusion-kinetic area.

At the alkali etching of fiber occurring from the surface and not complicated by the Rebinder effect, i.e. after completion of the chemosorption processes, the reduction of the volume component of the strength force is described, as it was shown in [11-13], by the following equation:

$$\frac{P_v(t)}{P_o} = \frac{P_v(0)}{P_o}(1 - k_v \cdot t)^2 \tag{6}$$

where: $P_v(0)/P_0$ is the relative value of the volume component at $t=0$, k_v is the effective constant of the volume component breaking force reduction due to the fiber etching, which is determined by the following expression:

$$k_v = k_r\,[NaOH]\,0.5d \tag{7}$$

where: k_r, $[NaOH]$ and d are the reaction basalt-alkali rate constant, the concentration of the alkali solution, and the average diameter of the elementary basalt fiber (monofilaments), correspondingly. Hence, the reduction of the breaking force by taking into account the simultaneously occurring processes of adsorption and fiber dissolution can be described by the equation, which is the combination of eqns. (4) and (6):

$$\frac{P}{P_o} = \frac{P_o - P_v(0)}{P_o} \cdot \exp(-k_s \cdot t) + \frac{P_v(0)}{P_o}(1 - k_v \cdot t)^2 \tag{8}$$

To determine the parameters $P_v(0)/P_0$ and k_v, the equation (8) can be transformed into the following one:

$$\left(\frac{P}{P_o}\right)^{0,5} = \left[\frac{P_o - P_v(0)}{P_o} \cdot \exp(-k_s \cdot t) + \frac{P_v(0)}{P_o}(1 - k_v \cdot t)^2\right]^{0,5} \tag{9}$$

At large enough time, $t > t_2$, when $exp(-k_s \cdot t) \to 0$, the expression (9) becomes as follows:

$$\left(\frac{P}{P_o}\right)^{0,5} \cong \left(\frac{P_v(0)}{P_0}\right)^{0,5} (1 - k_v \cdot t) \quad (10)$$

According to (10), the reduction of the breaking force at large enough t (in the region 3 at $t > t_2$), in the coordinates $(P/P_0)^{0.5}$ - t, will be described by the straight line, which intercept with ordinate axis and give the part equal to $(P_v(0)/P_0)^{0.5}$. From the tangent of the slope angle of the line, it is possible to determine the constant k_v. Indeed, as it is seen from Fig.2, the experimental data can be satisfactorily described in the coordinates of equation (10) by a straight line with the parameters $(P_v(0)/P_0)^{0.5} \approx 0.85$ and $k_v = 3 \cdot 10^{-3}$ (days)$^{-1}$. The values of the kinetic parameters of Eqn.(8) of the basaltic roving aging in the NaOH solution are given in Table 2.

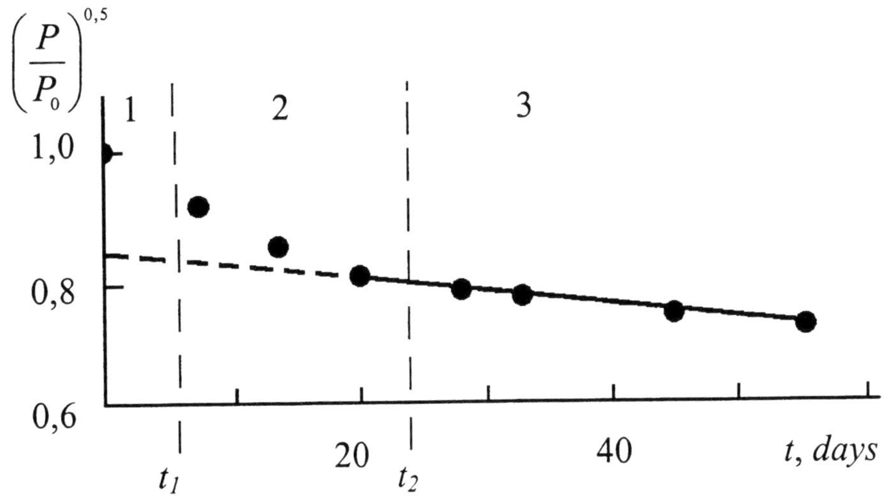

Figure 2. Dependence of the relative breaking force of the basaltic roving on the time of its aging in the NaOH aqueous solution (1.25N) given in the coordinates of Eqn. (10). The dashed lines are the borders of the regions 1, 2, and 3.

Table 2. Kinetic Parameters of the Basaltic Roving Aging in the 1.25N Aqueous NaOH Solution.

k_s	k_v	k_{vm}	k_r	k_{rm}	$\frac{P_v(0)}{P_0}$ *	$\frac{P_s(0)}{P_0}$ **
$\frac{1.25\ 10^{-1}}{1.45\ 10^{-6}}$	$\frac{3.0\ 10^{-3}}{3.47\ 10^{-8}}$	$\frac{1.25\ 10^{-3}}{1.45\ 10^{-8}}$	$\frac{5.33}{6.17\ 10^{-5}}$	$\frac{2.22}{2.57\ 10^{-5}}$	0.72	0.28

Notes. The dimension of the parameters k_s, k_v, and k_{vm} $(days)^{-1}$ (nominator) and s^{-1} (denominator); k_r and k_{rm} are the effective constants of the alkaline etching of the basaltic fibers, they were calculated according to Eqn.(7) based on the data of mechanical testing and on the mass loss, correspondingly. The fiber diameter 9 μm. The dimension of the constants k_r and k_{rm} is $(days)^{-1}\cdot cm^{-1}/(mole/l)$ (in the nominator) and $s^{-1}\cdot cm^{-1}/(mole/l)$ (denominator).

* - Volume component of the breaking force, ** - Surface component of the breaking force.

As it is seen from Fig.2, the calculations (line) performed according to Eqn.(8) and the obtained parameters (Table 2) are in a good agreement with the experimental data (points) in the investigated interval of the samples aging time change.

Based on the found value of k_v, and formulas (8)-(10), it is possible to evaluate the maximal limited time

$$t_{lim} = 1/k_v \tag{11}$$

of the samples alkaline aging, for which the decrease of the breaking force approaches zero. According to equation (11), the calculations for the investigated basalt fiber bundle gave the $t_{lim} \approx 330$ days.

The constant k_v for the fiber aging is the effective characteristic of the chemical stability of the fiber having a certain diameter in the alkali solution of a certain concentration. In other words, it is a technical characteristic of an aging of a concrete fiber material at given conditions. For a more general characteristic of the reaction ability of basalt toward to the alkaline solutions, the constant k_r can be used (Table 2). The constant k_r is numerically equal to k_v in case the fiber has a single diameter and a single unit solution concentration.

It is known that in a general case of the uniform surface etching and fiber dissolution, and in the absence of the clearly defined effect of the adsorption decrease of the breaking force, the following expression is fulfilled [11, 12]:

$$\left(\frac{m}{m_o}\right)^{0,5} = \left(\frac{F(t)}{F_0}\right)^{0,5} = \left(\frac{P(t)}{P_0}\right)^{0,5} = (1 - k_v \cdot t) \tag{12}$$

where: $F(t)$ and $F(0)$ are the current and the initial value of the fiber cross-section area, correspondingly. By representing the data on the mass loss (Fig.la) in the coordinates of the Eqn. (12) (Fig.3), the value of the k_{vm} was determined based on the obtained line slope angle tangent in the region $t > t_2$. It was found to be $1.25\cdot 10^{-3}$ $(days)^{-1}$. The comparison of the etching effective constants, obtained from the data on the breaking force reduction, k_v, and from the data on the mass loss, k_{vm} (Table 2), showed their significant difference. The found differences are related, basically, with the not complete dissolution of basalt in the outer aged fiber layer. At the gravimetric analysis of the aged fibers, the decrease of their mass was found to be only due to the dissolved basalt components. The non-dissolved part of the fiber aged surface layer in this analysis method is considered as the not aged part from the formally-kinetic point of view. This leads to a formal lowering of thefiber aikali aging efficiency.

In the deviation from the gravimetric measurements, at the mechanical testing, the fiber aged surface layer, having the poorly packed structure, does not significantly affect the measuring breaking force. In this case, the breaking force is determined only by the fiber inner trunk diameter, which was not touched by the aging. Hence, from the mechanical testing data, which are not sensitive to the existence of the fiber insoluble part in the weakened aged surface layer, the only effective constant, $\boldsymbol{k_v}$, is found. The constant formally corresponds to the completely dissolved (absent) outer aged surface layer. This can explain the higher value of $\boldsymbol{k_v}$ over $\boldsymbol{k_{vm}}$ (and $\boldsymbol{k_r}$ over $\boldsymbol{k_{rm}}$ correspondingly).

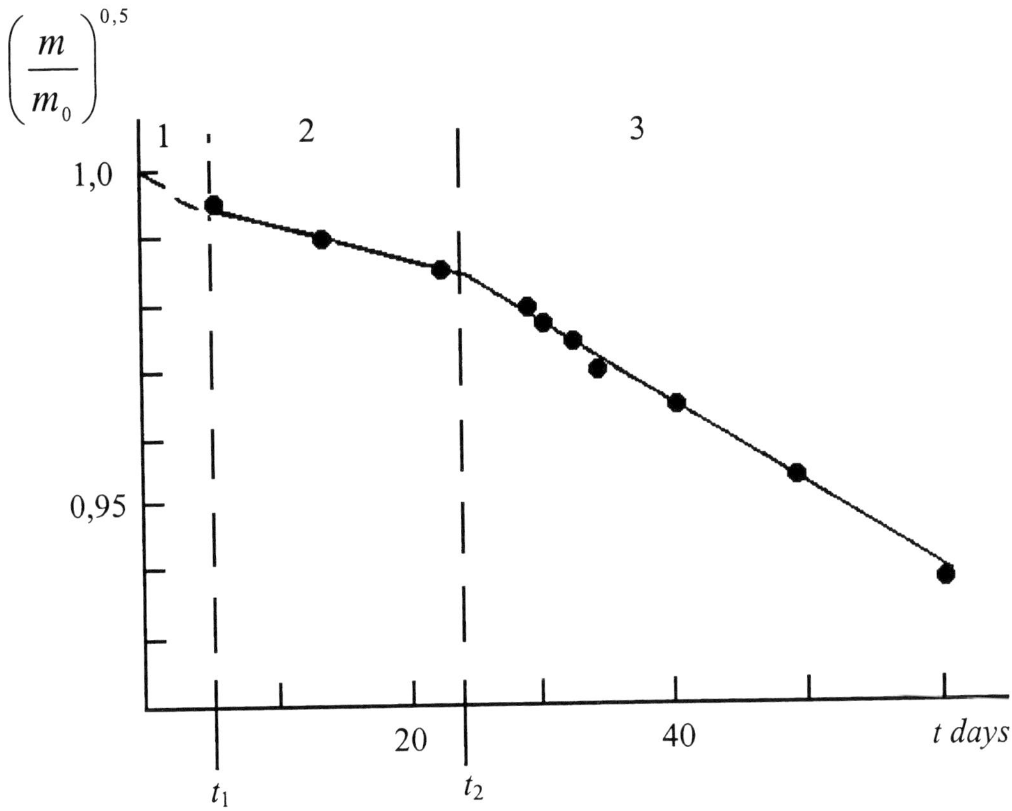

Figure 3. Dependence of the relative mass of the basalt roving on the time of its aging in the alkaline media (1.25 N solution of NaOH), which is represented in the coordinates of the Eqn. (12). The dashed lines are the borders of the regions 1, 2, and 3.

CONCLUSIONS

In this work, the kinetic regularities of the breaking force reduction and the mass loss at the basalt roving aging in the alkaline media have been investigated, and the parameters of the regularities have been determined. The obtained equations are in good agreement with the experiment. The equations take into account the reduction of the basaltic fiber breaking force caused by the occurrence of the adsorption processes and the partial chemical dissolution of the surface layer basalt components in the aqueous alkali solution.

The obtained regularities can be used for the prognosis of the tensile strength reduction of roving and other basaltic fiber materials under prolonged action of the alkaline aggressive media.

REFERENCES

[1] Trostyunskayu E.B., Kutyrev Yu. V. //*Plast Massy*, 1976, No.ll, pp.44-46.
[2] Troslycmskuyci E.B., el al., //*Plast Massy*, 1987, No.l, pp.28-31.
[3] Pashchenko A.A., Serhin V.P., "*Reinforcement of cement stone by the mineral fiber*". Kiev, UkrNIINTI, 1970, pp. 18-44.
[4] Aslanova M.S., Kolesov Yu.L, Khazanov V.E., Khodakovsky M.D., Sheiko V.E. "*Glass fibers*", Ed. Aslanova M.S., M. Khimiya, 1979, 1986, pp.155-180.
[5] Kalinichev V.A., Makarov M.S., "*Wound glass plastics*", M. Khimiya, 1986, pp. 19-43.
[6] Bulanov l.M., Vorobey V. V., "*Technology of missile and air-cosmic constructions made of composite materials*" (Russ.), M. N.E. Bowman MGTU, 1998, p. 18.
[7] Gromkov B.K., Smirnov L.N., Trofimov A.N., Zharov A.I., Semenov O.N., Yudin A.A., Koshelev V.Yu. //*Conversion in machine design*, 2000, No.4, pp.81-86.
[8] ZackA.F., "*Physico-chemical properties ofglass fibers*" (Russ.), Rostechizdat, 1962. pp. 138-178.
[9] Gunyaev G.M., "*Structure and properties of polymeric fiber materials*" (Russ.), M. Khimiya, 1981, p. 40.
[10] FrolovYu.G., "*Course of colloid chemistry*", M., Khimiya, 1981, p.40.
[11] Moiseev Yu.V., Zaikov G.E., "*Chemical stability of polymers in aggressive media*" (Russ.), M., Khimiya, 1977. pp. 162-165.
[12] Perepelkin K.E.//*Plust Massy.* 1977, No. 10, pp. 36-39.
[13] Perepelkin K.E., "*Structure and properties of fibers*" (Russ.), M., Khimiya, 1985, p. 144.

Chapter 6

RECICLABILITY OF POLYSTYRENE IN SOLID WASTES

***Juan López*[1], *Rafael Balart*[1], *Lourdes Sánchez*[1] *and Alfonso Jiménez*[2]**

1. Department of Mechanical and Materials Engineering. Polytechnic University of Valencia in Alcoy. Alcoy. 03801. Spain. e-mail: jlopezm@mcm.upv.es
2. University of Alicante. Department of Analytical Chemistry. PO Box 99, 03080, Alicante, Spain. e-mail: alfjimenez@ua.es

ABSTRACT

Recovery of styrenic materials from solid urban wastes by mechanical recycling has been studied. The main goal of this work was to know the possibility of recovering the material in order to get a competitive product similar to the virgin resin. The work starts with a review of two of the main problems for recovery processes, i.e. the elimination of impurities of other plastic materials, such as polypropylene and PVC, and degradation produced by previous recycling processes or the action of external agents such as radiation.

It is to note that the recovered material shows a clear decrease in mechanical properties as well as produces serious difficulties in the recovery process. Some analytical techniques, such as Differential Scanning Calorimetry can be used to predict the quality of the recovered material taking into account the main characteristics of the residue.

Keywords: Polystyrene; Recycling; Degradation; Mechanical properties; Calorimetry

INTRODUCTION

Polymeric materials present in Urban Solid Wastes are difficult to separate and recover by their intrinsic characteristics as well as the large variety of different materials that can be thrown away in modern life. In spite of not representing more than 7.3% in weight, plastic residues cover near 20% in volume because their low density.

Among all kinds of plastics present in Urban Solid Wastes, styrenic materials, mainly polystyrene (PS), represent the third larger amount, only after polyolefins and PVC, with approximate quantities of 7.5% of the total residue (1). This group of materials is highly heterogeneous, with commodities such as PS and technical polymers, such as styrene-acrylonitrile copolymers (SAN) and butadiene-styrene (ABS).

But, it is not easy to get pure PS in wastes, as it is usually obtained together to other chemically similar materials, such as polyamides, polyesters and PVC. These materials are difficult to isolate form PS, representing a serious problem in PS recovery. Another problem associated to PS recycling is the variation of one parameter called reciclability depending on the origin of residues. While in metals there is no influence of the material origin and they can not be clearly separated from the ore, this is not the case in PS recycling.

The main results in the characterization of the recovered materials is the main goal of this work, with proposal of solutions to get a high quality material to permit its further reuse. The characterization will be focused in composition and degradation processes, which are directly dependent on the residue origin.

Recovery of Residues from Domestic Wastes

Municipal solid waste (MSW) is one of the main origins of plastic residues. The European Commission marked in legislation highly strict conditions for the treatment of residues (94/62/CE Directive). They are usually thrown to landfill and the main objective proposed in legislation was to get recycling amounts of 25% of the whole material.

Among all plastics present in MSW, the main fraction corresponds to polyolefins, (polyethylene and polypropylene), and can be separated by flotation, with separations higher than 50% (2). The remaining fraction corresponds to a mixture of styrenic materials with other non-compatible polymer materials, such as PVC and PET. These polymers do not present any compatibility and low possibilities of recovery. Solution with trichloroethane and further reprecipitation with methanol can separate the styrenic fraction. But these solvents are expensive and therefore separation of styrenic materials of PVC and PET is scarcely used nowadays.

Another possibility of separation can be the mixing by extrusion of both materials with formation of a blend mainly constituted by PS and PVC. However, as PVC is easily affected by thermal degradation, extrusion conditions must be very soft, as indicated in Table 1 (3).

Table 1. Extrusion Conditions for PVC/PS Blends

Extrusion Stage	Temperature (ºC)
Screw (1)	130
Screw (2)	140
Filter	140
Exit	150

This operation permits the processing of pellets for further injection. However, the low compatibility of the original resins and the great variety of origin of both materials conditions the low regularity in mechanical properties of blends (1, 2). This is caused by the low mixture of PS and PVC that does not improve in any sense the original properties of virgin materials as well as the observed delamination of the blend. (Figures 1-2).

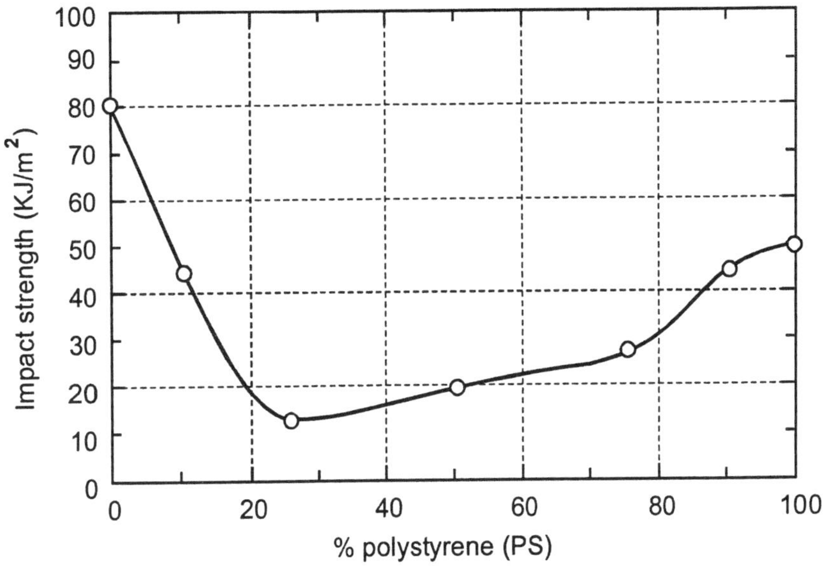

Figure 1. Variation of impact properties with the amount of PS in the blend

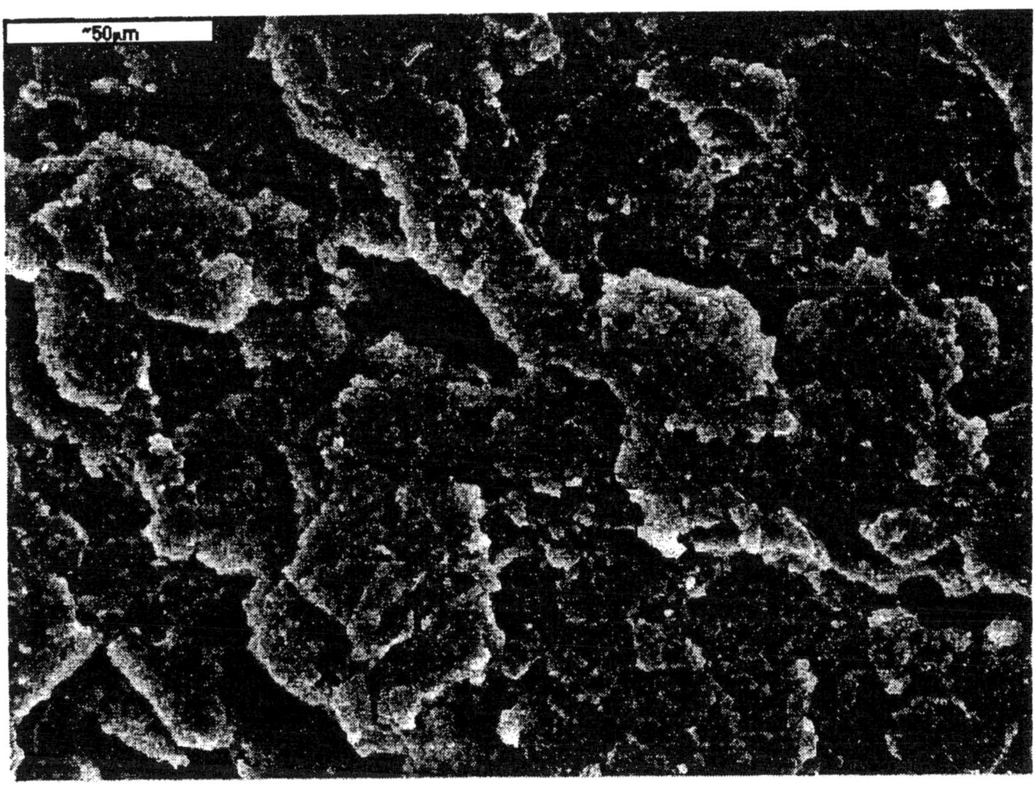

Figure 2. Delamination of non-compatible blends in styrenic materials

Therefore, a residue with styrene materials and vinyl chloride derivatives will cause problems in the recycling process, as well as low compatibility and bad mechanical properties. Thus, it is non-acceptable a direct recovery of such materials and a previous separation is necessary in order to get both polymers. The use of compatibility agents could permit a slight improvement in the performance of the blend. This can be carried out by the use of blending reagents (3-6) and more economically, plasticizers. Thus, the addition of diethyl-hexyl phthalate (DEHP) improves the blend properties, as it can act as a compatibilizer as well as lubricant. The addition of 30% DEHP or the use of plasticized PVC (P-PVC) can improve the compatibility and in particular the impact properties.

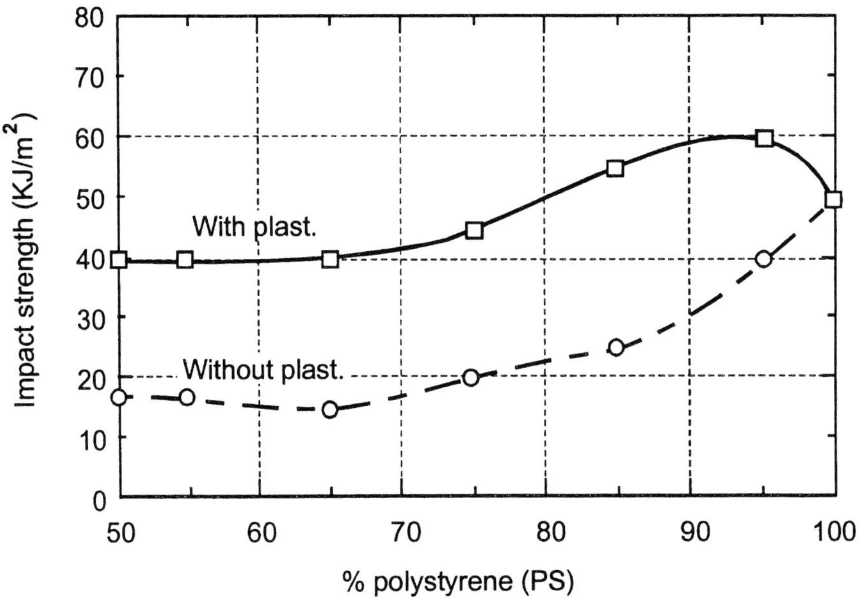

Figure 3. Variation of breakdown energy for PS/PVC blends with and without plasticizer

However, the addition of plasticizers does not improve the general properties of blends to make the process economically viable from the industrial point of view. Therefore, the selective separation and the processing of homogeneous materials is the main condition for further recycling of materials.

INDUSTRIAL RESIDUES OF POLYSTYRENE AND ELIMINATION OF IMPURITIES

Recovery of industrial residues is currently one of the main fields of effort by private and public institutions (7-8). The treatment of PS residues is not easy because its intrinsic structure, which is quite influenced by impurities, as well as degradation produced by external agents. The use of butadiene as impact modifier is useful for some PS applications but is an added difficulty for the recycling process.

It is usual in recycling of industrial residues to make a difference between injection and extrusion materials. The first group corresponds to residues from electrical wires and other household products. The main problem for recycling of these products is the presence of other plastic materials that make difficult the whole process and influence the degradation. The most usual contaminants in PS and SB samples are the following polymers:

- Methylene polyoxyde (POM) in electrical and electronic residues.
- PP in injection residues from household products.
- Polyesters in food industry residues.

POM is one of the most important contaminants in the PS recovery as it is used in many electronic devices. It is easily detectable by common analytical techniques, such as infrared spectroscopy (Figure 4) and thermal analysis (Figure 5). The presence of POM not only affects the recovery process but also influences the mechanical properties of the recycled product as indicated in Table 2. It can be observed that mechanical properties suffer a general decrease with the addition of 5% POM to a PS sample.

Table 2. Mechanical Properties of Pure PS and PS Blended with POM (5%)

Property	PS	PS + POM (5%)
Young Modulus (N/mm^2)	1255.38	1192.22
Tensile strength (N/mm^2)	21.83	17.72
Elongation at break (%)	12.19	4.66

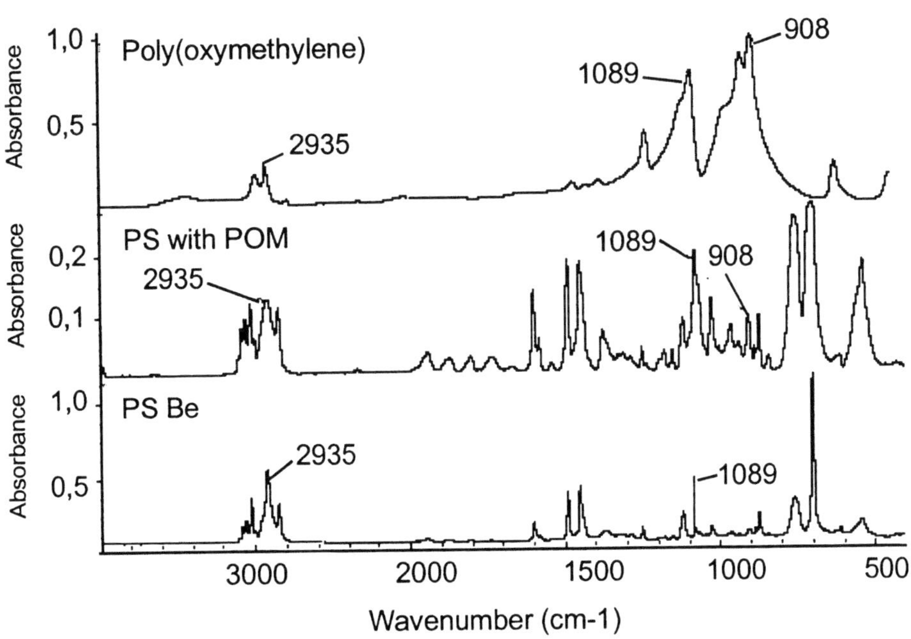

Figure 4. IR spectra of POM (a), recycled PS + 5% POM (b) and recycled PS (c)

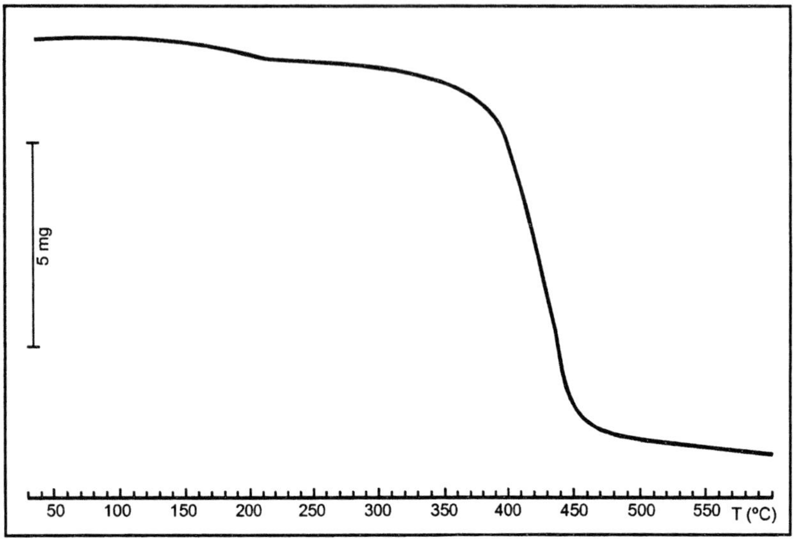

Figure 5. Determination of thermal degradation of POM in recycled PS

Polypropylene is the main contaminant in PS recovery because it is found in many different samples. The main effect of PP is the acceleration of PS degradation and therefore it makes the processing difficult. PP has an oxidation rate much higher than polyethylene (PE) and it effect is much more negative in PS recovery. This effect is especially important not only for the extrusion but for further processing and use, as observed in the calorimetric analysis shown in Figures 6a and 6b. After the first extrusion both materials show a similar degradation behavior, but after injection and other processing conditions, the material containing PP is thermally less stable.

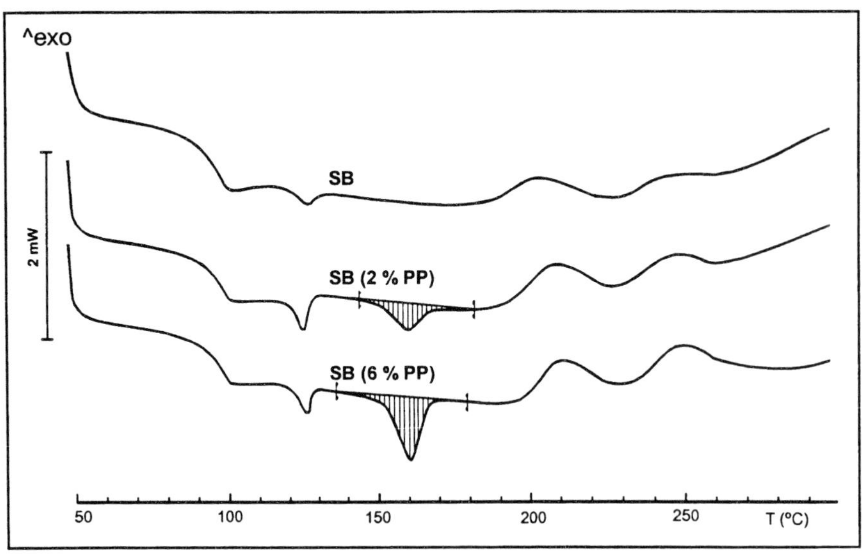

Figure 6a. Calorimetric curves of SB recovered and contaminated with PP (2 and 6 %)

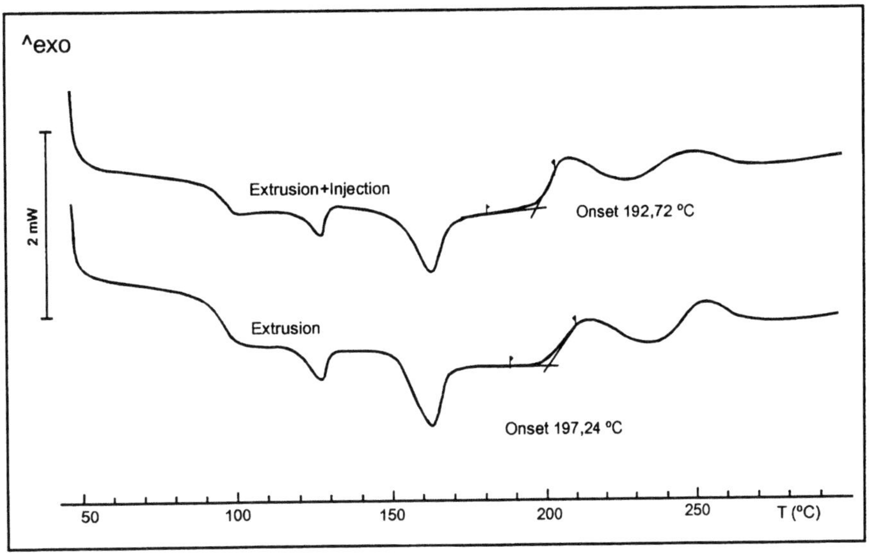

Figure 6b. Thermal degradation of recycled SB after one extrusion and one injection after extrusion

Polyester is the third contaminant in recycled PS, which is found in many samples, particularly in packaging of milky products. In this case, residues of packaging materials are grounded and prepared to get an homogeneous residue, which is further treated with water and NaOH to permit solution of aluminum also used in this kind of packages. Another possibility is the use of a mechanical system for the separation of polyester and PS before the first extrusion.

Table 3 shows the amount of different compositions of three milky products packages formed by PS, paper, aluminum and polyester.

Table 3. Composition of Three Different Milky Products Packages

	PS (%)	Paper (%)	Aluminum (%)	Polyester (%)
1	82.4	16.2	0.5	0.9
2	79.6	17.8	0.0	0.6
3	82.1	16.4	0.6	0.9

Grounding is the most important process in PS recovery in this kind of products. SB fragments for recovering and short polyester layers constitute the grounded material. Therefore, a clear difference in the weight distribution of both fractions can be observed. Most of the PS fraction shows weight between 10 and 40 mg, while polyester shows layers from 1 to 3 mg. This behavior can be observed in Figure 7.

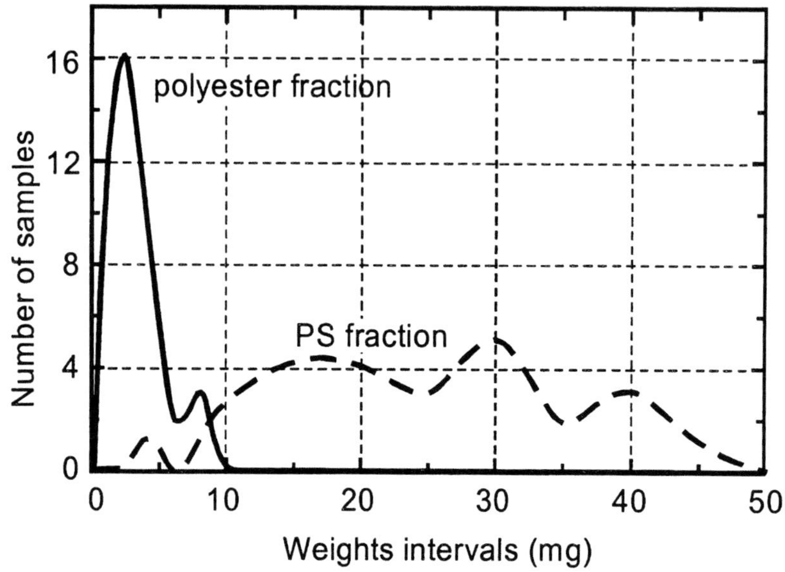

Figure 7. SB and polyester weight distribution in the grounded material

Two additional problems can be attributed to the presence of polyester in PS recycling.

- Formation of undesired gases by vaporization during the processing stage if it is not eliminated before extrusion. This can lead to the formation of impurities in the processed material. A DSC curve for polyester shows one peak near 100ºC attributed to the water adsorbed in the exterior layer of the material (Figure 8)
- The hydrolysis of the material after reaction with water producing undesired alcohol and acid molecules.

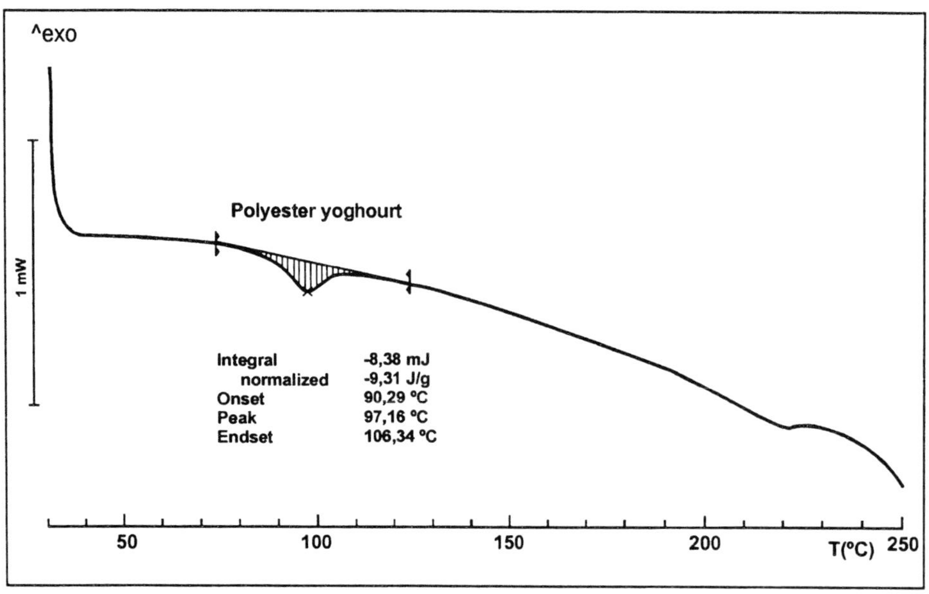

Figure 8. DSC curve of polyester used in packaging materials

OXIDATION AND THERMAL DEGRADATION

Thermooxidative stability of non-modified PS is relatively high. This is the reason why this material can be used for many applications with no need of additives in its formulations. However, repetitive processing, as in the case of recycling and reuse of the material can lead to some degree of degradation and further oxidation. In this case, damage can be observed by a general decrease of the molecular weight of polymers making them useless for some applications. A secondary effect of this decrease in the molecular weight is a significant increase in Melt Flow Index and a general decrease in mechanical properties (Figure 9).

Photooxidation is another important factor for the thermal degradation of this material. For instance, in High Impact Polystyrene copolymers (PS-HI), thermal energy caused by hydroperoxides degradation is responsible of the photooxidative degradation of the material. Rubber phases, such as polybutadiene, are highly sensitive to oxidation and therefore responsible of photooxidation of many copolymers.

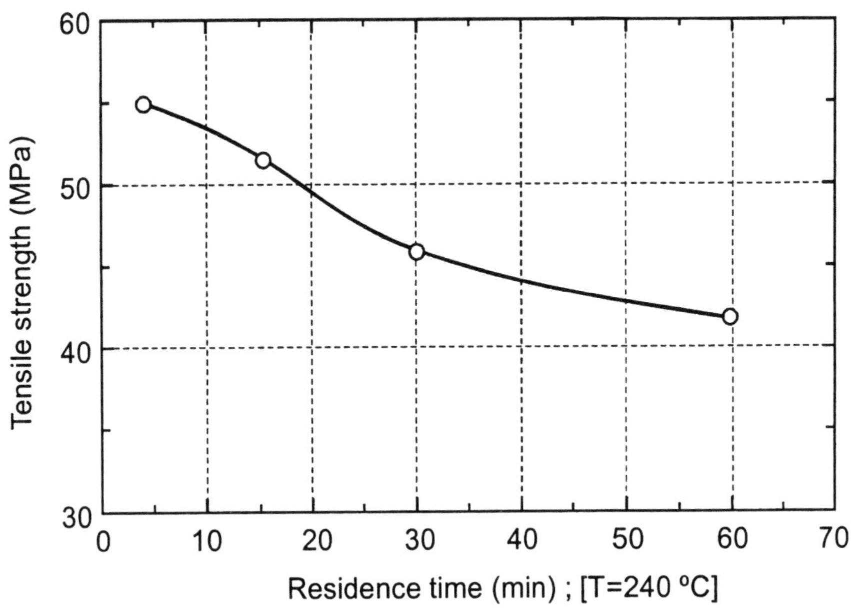

Figure 9. Variation of tensile strength with time of residence in machine

In the case of PS modified with butadiene forming short layers, the process is autocatalytic and an important decrease (higher than 50%) can be observed only after 5 weeks of exposition to sunlight (Figure 10).

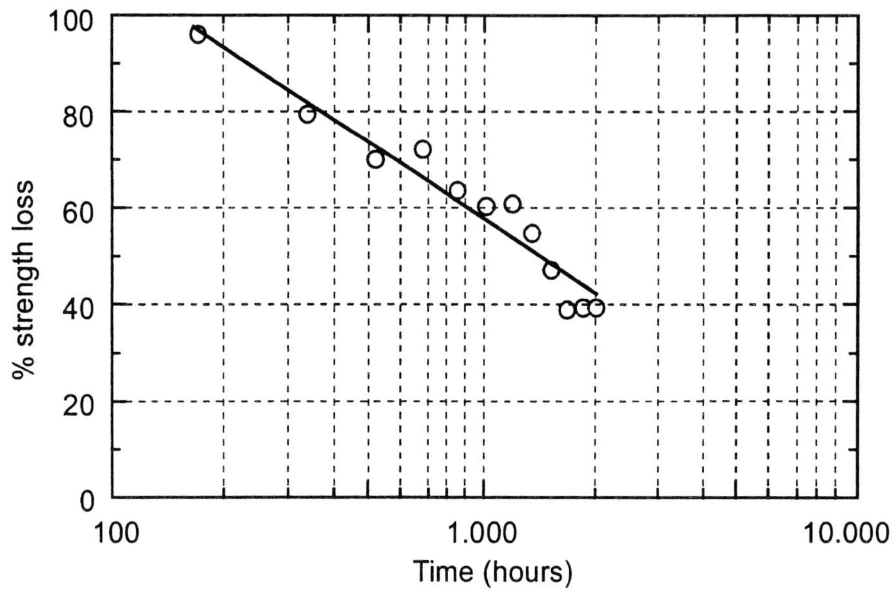

Figure 10. Loss of mechanical properties for PS-SB layers (2mm. thickness) with sunlight

This general loss of properties follows an exponential behavior, as indicated in Equation 1, where the degradation from the external to the inner part of the material can be explained.

$$\partial X = F\left[\left(1-\exp(-Kt)\right)^2\right] \quad (1)$$

The second effect of this degradation is the bad reaction to the recycling process caused for the effect of sunlight radiation as observed in Figure 11.

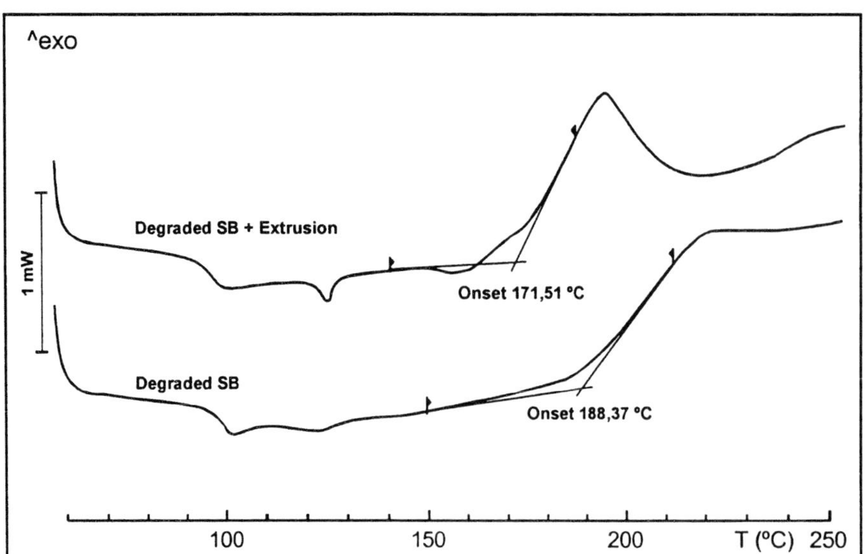

Figure 11. Degradation of a material exposed to sunlight

Therefore, it is to note that the SB component shows a strong sensitivity to radiation. These results show a clear dependence between the butadiene content, which is degraded in function of time, and the rate and mechanism of photodegradation. The complete failure of the material is only question of the exposition time of the material.

CONCLUSIONS

The recycling of styrenic materials shows a great interest from the economical and environmental point of view. However, and showing a difference to other common polymers, such as polyolefins, it shows some problems caused by several factors. The recovery from domestic wastes, with no previous preparation is practically non-viable as the difficulty to find systems to get an efficient and economically viable separation.

In the case of industrial residues the process is possible but it is still subjected to problems caused by the intrinsic structure of styrenic copolymers. The main problems are the presence of impurities from other polymers, i.e. POM, PP and polyesters. The origin of residues and the thermal history of materials are also of great importance as well as photooxidation to the possibility of an efficient recycling process.

REFERENCES

[1] Ambrose & al, Resource, *Conservation and Recycling* **36**, 309-318 (2002)
[2] Villalba G., & al. Resource, *Conservation and Recycling* **36** (2002) in press
[3] Chuai C., Almdal K.,Johannsen I., Lyngaae, *Polymer Engineering & Science*, **42**, 961-968 (2002)
[4] Eguiazabal J., Nazabal J, *Polymer Engineering & Science*, **30**, 527-531 (1990)
[5] Keitz, Barlow, Paul, *J. Appl. Polym. Sci*, **29**, 3131-3145 (1984)
[6] Ohishi H, Ikehara T, Nishi T., *J. Appl. Polym. Sci*, **80**, 2347-2360 (2001)
[7] M.Diaz, S. Barbosa, *N. Capiati. Proceedings of Archipol* 2001, 195
[8] Alberti J., Bullach R.L., Ellis T.S. & McMinn, *Polymer Recycling* **2**, 159-171, 1996
[9] K.Z. Gumargalieva, G.E. Zaikov, in *Polymer Analysis Degradation,* A. Jiménez, G. E. Zaikov (Eds), Nova Science Pubs. (2000) p 1

Chapter 7

HETEROGENEOUS PECULIARITIES OF POLYMERIC HYDROCARBONS' AUTOOXIDATION

Yu. A. Mikheev, G. E. Zaikov and V. G. Zaikov
N.N.Emanuel Institute of Biochemical Physics
Russian Academy of Sciences
Kosygin st. 4, Moscow, 117977 Russia

ABSTRACT

The advantage of colloidal-heterogeneous reactionary model of autooxidation of non-crystalline polymeric hydrocarbons versus homogeneous-liquid phase one of the process is demonstrated with a kinetic analysis. The model proceeds from character of supermolecular organization of polymeric chains for a type of spongy micelles, including a non-uniform porous zone of nanostructural scale. The mechanism of oxidation of micellar hydrocarbon sponge takes in account distribution of the conjugated reactionary chains on non-uniform nanophases, distinguished on the sizes and thermofluctuational dynamics of its nanopores. Kinetic equations of macroscopic stages of the process are given.

Keywords: Micella, sponge, kinetics, oxidation, heterophaseous, hydrocarbons, non-crystalline, pore, macroscopic, spherolite

1. INTRODUCTION

For many years process of oxidation of block polymers is treated in the context of the reactionary schemes of homogeneous oxidation of liquid hydrocarbons, not reflecting specifics of supermolecular organization of chain molecules [1 - 13]. The formal substantiation for this is an existing for along time sight on homogeneous amorphism of a non-crystalline polymeric phase. The obvious physical difference of the latter from usual liquids is simplifiedly connected to unstable formations, functioning in it, which continuously appearing and breaking up, but distinguished by rather long lifetime and low rates of the relaxational response to external influences [14].

Meantime, as a rule, kinetic regularities of the polymer oxidation do not satisfy in current frameworks of homogeneous reaction. It is the reason of occurrence of many ambiguities upon consideration of mechanisms of such macroscopic stages, as an induction period [5] and autoacceleration. The latter is related to two different mathematical dependences: a). so called «parabolic» kinetic law; b). exponential law *exp(Φt)* [4 - 6, 8 - 13, 15]. The contradictions caused by model of homogeneous reaction bear doubts [16] of mechanisms of inhibited oxidation and "critical" concentrations of antioxidants, formulated in the literature [4, 6 - 8].

The schemes inadequate to molecularly organized systems do not take in account the influences of a real physical structure of a polymeric matrix on the chemistry of oxidation and do not reveal mathematical interrelation between kinetics of oxygen's absorption and kinetics of polymeric hydroperoxide transformation. Bearing on them upon examination of different facts, the researchers quite often have to introduce mathematical simplifications and postulates, creating the contradictions either with already earlier used postulates, or with the experimental facts, changing thus contents of schemes and physical essence of the kinetic equations [1 - 13, 15, 16].

As an example we shall result one of last publications [17], where the author, operating with model of homogeneous reaction, offers to consider a rate of accumulation of hydroperoxide on all stages of developed oxidation of polyolefin equals zero:

$$d[POOH]/dt = \alpha V - k_{hp}[PH][POOH] = 0 \tag{1}$$

where V - rate of oxygen's absorption, α - yield of polymeric hydroperoxide at absorption of one gramm-mol of oxygen, [*PH*] and [*POOH*] - concentration of reacting bonds CH and hydroperoxide groups respectively, k_{hp} – rate constant of decomposition of POOH in proposed reaction of initiation of radicals

$$POOH + PH \rightarrow (PO^{\bullet} + P^{\bullet} + H_2O) \rightarrow \sigma P^{\bullet}$$

where σ - coefficient of yield of free macroradicals $P^{\bullet}$.

Deduced on the basis of equation (1) ratio [17]

$$V = k_1[P^{\bullet}][O_2] = k_2[PH][POO^{\bullet}] = k_e[PH]^2 = k_{hp}[PH][POOH]/\alpha$$

requires synchronous change of concentration of radicals $POO^{\bullet}$, groups PH, absorbed oxygen (N) and accumulating peroxide:

$$k_2[POO^{\bullet}] = k_e[PH] = k_e(q[PH]_0 - N) = k_{hp}[POOH]/\alpha$$

where k_1, k_2, k_e - constants, q - number of moles of oxygen consumed for oxidation of one mole of PH [17].

Meantime, the experiments bear evidence about the unsynchronous change of POOH concentration relatively to [PH] and N. Actually absorption of oxygen and disappearance of PH-groups proceeds practically with constant rates on that interval of time of the developed process, when hydroperoxide is collected first, and then reacts after achievement of the maximal concentration [15, 18 - 22].

Let's note that the postulate [17] does not satisfy to reactionary model of homogeneous oxidation of hydrocarbons as well, because it requires direct proportionality between concentrations of $POO^{\bullet}$ and POOH instead of characteristic dependence $[POO^{\bullet}] \sim [POOH]^{0.5}$ for liquid phase process.

The fixed insight about liquid phase-homogeneous structure of a non-crystal polymeric matrix has not allowed to formulate the adequate mechanism of oxidation even when the dependence of process' rate from thermomechanical state of a polymeric matrix [4 - 6, 23 - 26] was established. The basic efforts of the researchers have concentrated on attempts to improve homogeneous reactionary model by attraction of ideas about polychromatism of kinetic parameters [5 - 7], presence of "centers" [4, 5] and "rigid cages" [8] in oxidizing matrix.

The model of kinetic "polychromatism" was created based on an idea that elementary (in homogeneous environments) reactions proceed in polymers as processes that do not have a certain rate constant. In such cases it is postulated that the reactivity of reacting particles depends on the nearest environment, reflecting somehow a character of organization or dynamics of the nearest parts of macromolecules. However the physical reasons of this phenomenon remain, as a rule, unknown.

The ideas about molecular-segmental "centers" and "cages" detail in the certain degree a picture of polymeric process. It is possible, that corresponding schemes are coordinated to separate concrete situations. At the same time, they can not be of the general basis for modeling reactions in a non-crystal polymeric matrix, because both "centers" and "rigid cages" are treated as parts of essentially homogeneous media. Modifying the same homogeneous reactionary model, all mentioned approaches do not describe the physical reasons, for example, why the oxidation of polymer in its qualitatively differing physical state quite often proceeds on the same kinetic law. For example, the temperature dependence of rate of uninhibited oxidation of low density poly(ethylene) (LDPE) in the induction period is described by the Arrhenius equation in an interval of temperatures 40-200 °C, that specifies an invariance of the process' mechanism at transition from a solid state through $T \approx 115$ °C in a melt [27].

The inconsistency of the homogeneous reactionary schemes (both classical and modified) is found out at discussion why the same polymer with its constant chemical structure of molecules frequently acts as qualitatively different reagents depending on a way of samples' preparation. Or why macroscopic stages with qualitative different kinetic laws (abnormal from the point of view of homogeneous reactionary model) are observed on different stages of its oxidation.

The reasons of noted kinetic peculiarities lie in real heterogeneity of "amorphous" polymeric phase, namely, heterogeneity of nanostructural scale, as an integral property of molecular-chain spongy aggregations, composing one of the main details of supermolecular skeleton [15, 16, 28, 29]. Molecular-chain sponge actually is the system of the connected "pendula" (oscillators), whose fluctuations have an opportunity to cooperate, forcing macromolecules' segments to be moved apart with formation of more or less large nanopores.

It is noted in a series of works dedicated to this topic [15, 16, 18 - 22] that initiating activity of the doped initiators and polymeric hydroperoxides, as well as the direction of free radical transformations, depends on a reagent's location in that or another structural nanoporous zone of a molecular-sponge matrix. Fine heterogeneity and the highly-elastic pliability to mechanical influences, characteristic for such matrix, provided significant variations of value of a stretching of polymeric chains, amplitude and frequency of pulsations of nanopores in its dimensions. Samples of the same polymer are able to react on qualitative unequal kinetic laws, despite of a same chemical structure of the molecules owing to difference in thermomechanical state depending on a way of preparation.

2. Influence of a Thermomechanical Condition of Polymer on Oxidation Kinetics

Internal thermomechanical condition of oxidizing polymer in many respects depends on character of those thermal and mechanical influences, which the polymer experiences before reaction. Nonequivalence of samples' background can result to qualitative unequal kinetic laws of their oxidation.

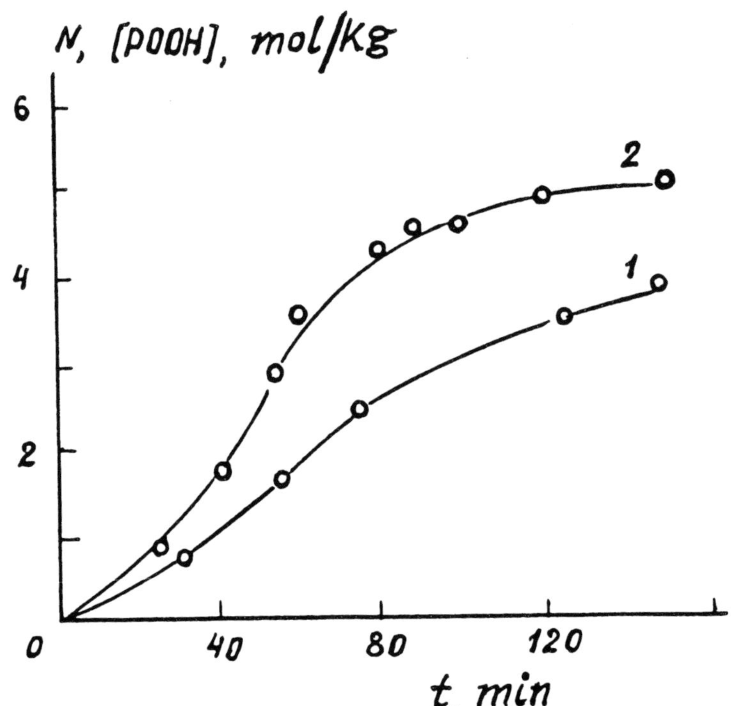

Figure 1. Change of concentration of hydroperoxide in films, pressed from unmelted powder IPP, during oxidation at 130 °C and at pressure of oxygen 26.7 (1), 53.3 kPa (2) [1, 2, 30].

This pertains, for example, to samples of isotactic poly(propylene) (IPP) in works of Pudov [1, 2, 30] and Monakhova [24, 31].

The curves 1 and 2 submitted on Fig. 1 show accumulation of POOH in Pudov's samples at 130 °C and pressure of oxygen p = 26.7 and 53.4 kPa respectively. They differ from the data of Monakhova (Fig. 2, curves 1, 2, the experimental points for a curve are summarized from works [24, 31], T = 130 °C, p = 40 kPa) by the duration of the process of accumulation POOH and by smaller rate of its decomposition. Really, according to Fig. 1, not less than 0.35 mol/kg of POOH is collected for 140 min of and the maximal concentration is not reached yet. Whereas according to Fig. 2 (curves 1, 2) maximal concentration of 0.25 mol/kg of POOH is observed after the induction period through 70-80 min, and through 140 min only about 0.15 mol/kg of hydroperoxide remains in the system.

It is established that confronted processes have qualitatively unequal dependences of absorption rates of oxygen on concentration of collecting POOH. Pudov correlates this dependence with sole macroscopic stage and is characterized by two close on value components:

$$V = a*[POOH]^{0.5} + b*[POOH]$$

where a, b - constants. Unlike this, the data of works [24, 31] bear evidence actually to three macroscopic stages of oxidation with unequal dependence V of [POOH]. Consecutive transition from more or less long induction period to a short stage of autoacceleration and then to a stage with practically constant rate of oxygen's absorption is observed on curves 3, 4 on Fig. 2 (they are given without experimental points, as well as in works [24, 31]). It is seen that at the third stage the concentration of POOH is increased first, but then is reduced during the certain time.

Let's note that Pudov prepared his films from a IPP powder with unmelted phase, subjecting a powder to short-term pressing (2 - 3 min) at temperature 125 - 130 °C [30] whereas in works [24, 31] an initial IPP powder was melted at 220 °C first. Cooling a melt under low pressure the films with fine spherolites were obtained, and after additional heating large spherolites were drown in them, but the degree of crystallinity did not change. The curves 1, 3 and 2, 4 on Fig. 2 related to fine and large spherolite films.

The feature of Pudov's films was that they reserved in themselves an internal mechanical tension connected with deformation of a skeleton of unmelted crystallites and with double-axial orientational stretching of polymeric chains, filling space between crystallites. Comparing curves 1, 2 on Fig. 1 and curves 1, 2 on Fig. 2 it is possible to conclude that the accumulation of POOH in an interval of concentration of 0 - 0.2 mol/kg in Pudov's samples proceeded almost with the same rate, as in films with fine spherolites. However at the subsequent stages of oxidation the yield of POOH is much higher at Pudov's films. The latter points at increasing of reaction rate in samples distinguished by internal tension. Apparently intensive relaxation of tensions with appearance of the local centers of thixotropic excitation of a matrix at the certain stage of process has occurred [32]. The frequency and amplitude of nanopores' pulsations of with change of the sizes and, respectively, rate of appearance (initiation) of free radicals [15 - 22, 33, 34] grew in these centers.

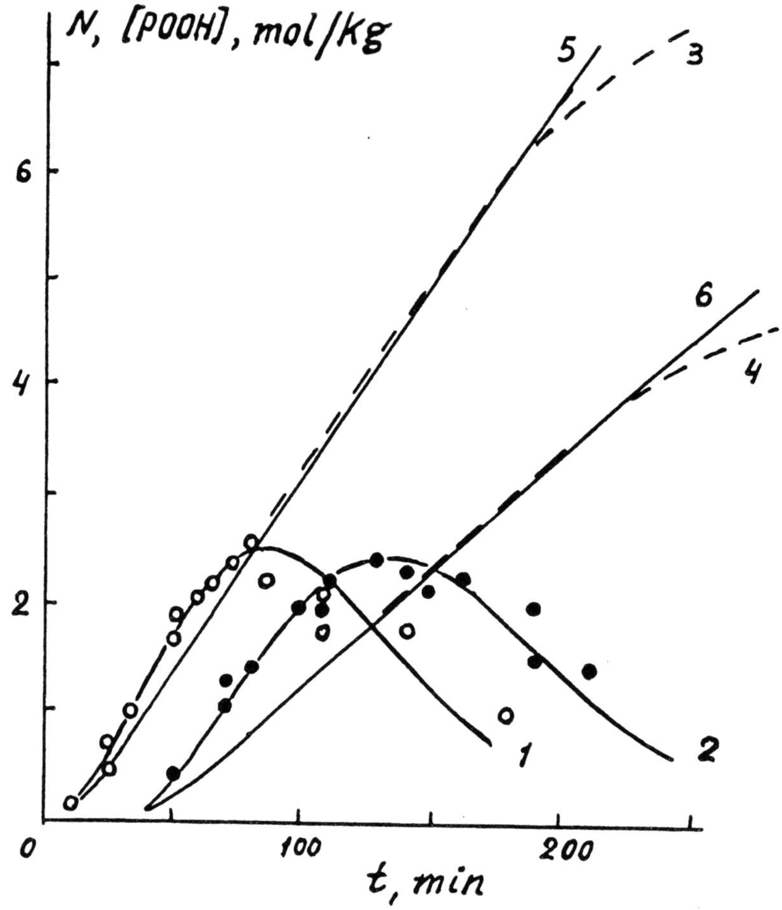

Figure 2. Calculated curves of hydroperoxide accumulation (1, 2), experimental (3, 4) and calculated (5, 6) curves of oxygen's absorption during oxidation of IPP films, prepared from melt and containing fine (1, 3, 5) and large spherolites (2, 4, 6). T = 130 °C, p = 40 kPa [24, 31].

IPP films, obtained by cooling melt under small pressure [24, 31], differed from the pressed conglomerate of not completely melted particles by much more equilibrated (in thermomechanical meaning) skeleton from forming crystallites. The residual internal tensions in them were considerably less, allocated more regularly in a material, and could not provide intensive thermomechanical excitation of a matrix and significant shear deformations during oxidation.

Observable increase of the induction period on Fig. 2 (curve 2) related with enlargement of spherolites as a result of their recrystallization can be explained that the part of macromolecules' segments is being built in crystallites, causing simultaneous orientational stretching of polymeric chains in areas between crystallites.

The stretched out polymeric chains are fixed strongly enough on crystallite skeleton and it leads to increased induction period.

The similar effect takes place at highly crystalline IPP films, high density poly (ethylene) and poly (methyl pentene), subjected to monoaxeal orientational stretching. Their induction periods of oxidation are increased that more, than more stretching degree is [5, 35].

Orientational stretching renders quite different influence on rate of oxidation of LDPE. As follows from work [35], its low degree of crystallinity does not provide sufficient resistance to internal tensions and quickly coming process or relaxation increases rate of oxidation from the very beginning of the process.

Thus, apparently, identical background of orientationally stretched films of polyolefins reflects differently on the initial reaction rate depending on durability of a supermolecular skeleton.

One more peculiarity of the oxidation of orientationally stretched IPP films consists in substantial growth of molecular weight of polymer on the initial stage, as opposed to intensive destruction of macromolecules in isotropic films [5]. Thus, as calculations have shown [35], the influence of stretch is caused not by change of solubility and coefficient of oxygen's diffusion, but by change of physical structure of oxidizing matrix IPP.

It is necessary to note that the consequences of stretch are observed at the autoaccelerated stage of oxidation as well. The letter becomes more long at the large degree of stretch [5, 26, 35] and the rate of oxygen's absorption in a number of cases is increased not proportionally of time, but according to the law $V \sim \exp(\Phi t)$ [26].

The state of a molecular-sponge matrix can change, and this affects on kinetics of oxidation, even as a result of simple heating of polymer. So, according to the data of work [20], oxidation of a powder of industrial unstabilized IPP on air at 120 °C combines with the steady induction period τ=3 h. However its heating in vacuum at 120 °C during 10 h has increased τ up to 5 - 7 h. The storage of thermotreated powder during 5 months in vacuum has not affected on the increased induction period, while storage on air at ambient temperature gradually decreased the value of τ (at 120 °C) up to 4 h in 1.5 months and up to 2 h in 4 months. The similar picture was observed with IPP samples, melted in vacuum at 200 °C. Their storage on air within 5 months has lowered τ up to 1 h instead of initial value of 5 h.

The chemical analysis of a IPP powder for the POOH contents has shown [20] that vacuum heat treatment at 120 °C (10 h) has not affected on initial concentration of [POOH] = $1 \cdot 10^{-4}$ mol/kg, however has caused a gain of POOH quantity up to $1.4 \cdot 10^{-3}$ mol/kg in a month of a storage on air. Thus, value τ was reduced up to 3 h, that equals a value τ of initial unheated powder. It turns out so that the thermal processing in vacuum reduces rate of IPP oxidation at raised temperature, increasing the induction period, but considerably increases rate of oxidation at low temperature (20 °C). Low temperature process was observed also with IPP films, prepared from melt and heated in addition in vacuum at 160 °C during 25 h: in 6 months the concentration of POOH in those samples has increased up to $3 \cdot 10^{-3}$ mol/kg [20].

Thus, changing the thermomechanical condition of a molecular-sponge matrix as a result of heating in vacuum, IPP more actively reacts with oxygen (probably under influence of moisture, absorbed from air [29]). Simultaneously with this there is also other, inexplicable within the framework of homogeneous reactionary model peculiarity, when the identical content of POOH in polymer bears different on value induction periods and, vice versa, the identical induction periods correlates with concentration of hydroperoxide, differing on the order of magnitude.

Let's note that a variety of the considered above regularities (observed in kinetical mode of oxygen's consumption) reflects general property of heterogeneous micro- and nanoporous bodies caused by lability of their physical structure. Such lability in many cases is undesirable, because it causes bad reproducibility of useful properties of materials [36].

Molecular sponge, being the important constructional material of a non-crystal polymeric matrix, has the strongly expressed pliability to thermomechnical influences and significant lability of its structure. Its properties essentially allocate non-crystal polymers from a file of homogeneous liquids, which do not change rates of its reactions at mechanical shear influences (mixing) and which, as a rule, differ unequivocal kinetical parameters of the elementary chemical acts.

3. Kinetic Peculiarities of Process in a Structural - Stationary State of a Polymeric Matrix

The concept of structural stationarity has some tentativity in relation to oxidizing polymer. The oxidation is inevitably connected to formation of oxygen containing groups on macromolecules as well as processes of cleave or networking of polymeric chains. All this changes initial balance of elastic forces in a molecular-sponge matrix, inducing more or less fast structural evolution. At the same time, if the chemical-mechanical disturbance is insignificant and matrix quickly restores a condition of thermomechanical balance, such process can be considered as structural–stationary one.

Structural-stationary processes of oxidation include several macroscopic stages, distinguished on its kinetic laws. So, the initial stage named in the literature as the induction period differs by low rate of oxygen absorption V_1. It ends after absorption of small amount of oxygen, usually less than 0.1 mol/kg. The value of V_1 does not depend of concentration of hydroperoxide, which is accumulated within induction period, but increases at elevated oxygen pressure, not reaching a limit according to expression [20] $V_1 = a_1 p/(b+p)^{0.5}$, where a_1, b - constants.

After an induction period there comes the short autoaccelerated stage of oxygen absorption, which rate (V_2) is increased more often directly proportionally of oxidation time. At this stage the reactionary chains are initiated by dissociated POOH, and perish according to the mechanism of square-law termination.

During oxidation the rate of a reaction rate's gain is slowed down, it reaches the maximal value (V_3) and ceases to depend on concentration of POOH. As against of V_1, the rate V_3 reaches its limiting value at increased oxygen pressure. It is essential, that the third stage of oxidation is established before the moment of the maximal concentration of hydroperoxide and remains the same for the long time even during significant subsequent decrease of concentration of POOH. Such dependence (observable, for example, on Fig. 2) is characteristic for zero order reaction on hydroperoxide [20 - 22] and essentially distinguishes bulk high-molecular weight polymers from their solutions.

Kinetic difference of processes in oxidizing bulk and dissolved LDPE is shown on Fig. 3 (experimental data of [37]). Curves 1 and 2 represent the dependence in melt (160 °C, p = 100 kPa), and curves 3 - 6 in a solution of polymer in trichlorobenzene (concentration is 0.56 mol/L calculated in account on monomeric repeated unit of LDPE, p = 75 kPa). The absorption of oxygen is characterized by curves 1 (in melt) and 3, 4 (in a solution at 160 and 170 °C, respectively). Kinetics of accumulation of POOH in both systems essentially differ (on Fig. 3 curves 2, 5, 6 show dependence of concentration of POOH of a doze of the absorbed oxygen) regardless of visual similarity of curves 1, 3, 4. As it is seen, the change of concentration of hydroperoxide in melt corresponds to a curve with a sharp maximum, while in a solution – to monotonous curves with rather low limits.

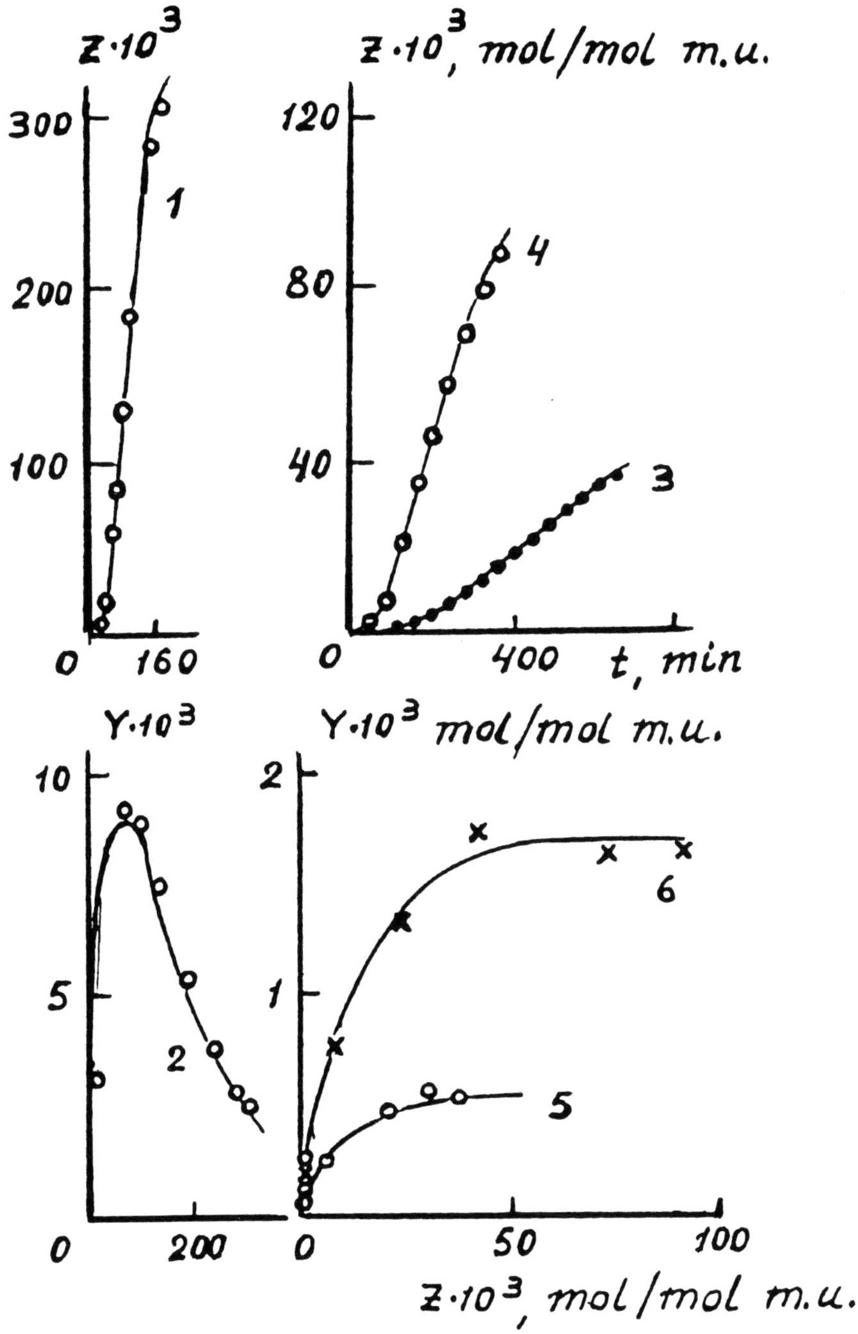

Figure 3. Dependence of quantity of the absorbed oxygen on oxidation time (1, 3, 4) and dependence of quantity of accumulated hydroperoxide from a doze of the absorbed oxygen (2, 5, 6) at oxidation of melt (1, 2; T = 160 °C, p = 100 kPa) and solution LDPE in trichlorobenzene (concentration – 0.56 mol/l, p = 75 kPa) at 160 (3, 5) and 170 °C (4, 6) [37].

Thus, according to the data of work [37], the reaction in a solution is described by the kinetic equation of liquid-phase oxidation

$$d[POOH]/dt = \alpha V - k_{hp}[POOH]$$

and reaches a stationary mode on those stages, when the CH-groups react rather insignificantly and when the conditions are provided

$$(d[POOH]/dt)_{\lim} = 0, \qquad \alpha V_{\lim} = k_{hp}[POOH]_{\lim} \approx const.$$

The oxidation of melt comes off on a stationary mode at an insignificant consumption of CH-groups (Fig. 3, curve 1), however change of concentration of POOH (Fig. 3, curve 2) does not reflect any dependence for liquid-phase reaction.

The following conclusions [15, 20 - 22] are made concerning oxidation of high-molecular weight bulk polyolefins: 1) hydroperoxide is not a unique primary molecular product of the reaction and other oxygen-containing compounds are formed in the early stages of the reaction, 2) decomposition of POOH in the advanced stage of transformation is caused by reaction with a product of oxidation and proceeds without formation of free radicals, 3) stationary stage of process of oxidation have zero kinetical order on hydroperoxide.

It is necessary to emphasize that zero order on hydroperoxide in the advanced reaction was repeatedly documented in many publications, but has not received the rational discussion (as it is seen, in the peak period of the theory of liquid-phase oxidation of hydrocarbons, the problem of structural - kinetical modeling has not ripened yet).

The doubtless importance of the given fact has required realization of its special verification employing a melt of poly(ethyleneoxide) (PEO) [15, 18, 19], highly-elastic copolymer of ethylene with propylene and rigid IPP [21]. The carried out experiments simultaneously have confirmed also another fact repeatedly documented earlier that presence of the doped initiator in the specified polymers, though accelerates an output of process on the maximal rate, but does not affect its value in a wide interval of the used concentration.

The results of experiments with PEO melt, oxidized on air [15] are submitted as an example on Fig. 4. Here curve 1 characterizes absorption of oxygen and, as it is seen, the constant value of V_3 is established before the maximal concentration of POOH (curve 2) is reached. The curves 3 and 4 correspond to a consumption of CH-groups of polymer and accumulation of aldehyde products of the reaction. They also confirm the fast achievement of constant rates, which are not dependent on the contents of POOH in melt.

Addition of the doped initiator – dibenzoilperoxide (BP) – to a PEO melt in concentration up to 0.12 mol/kg removes a stage of autoacceleration, but is not reflected on stationary rate of accumulation of aldehyde (Fig. 4, straight line 5, parallel to a straight line 4) and hydroperoxide (Fig. 4, straight lines 7 - 9, parallel to a straight part of a curve 6).

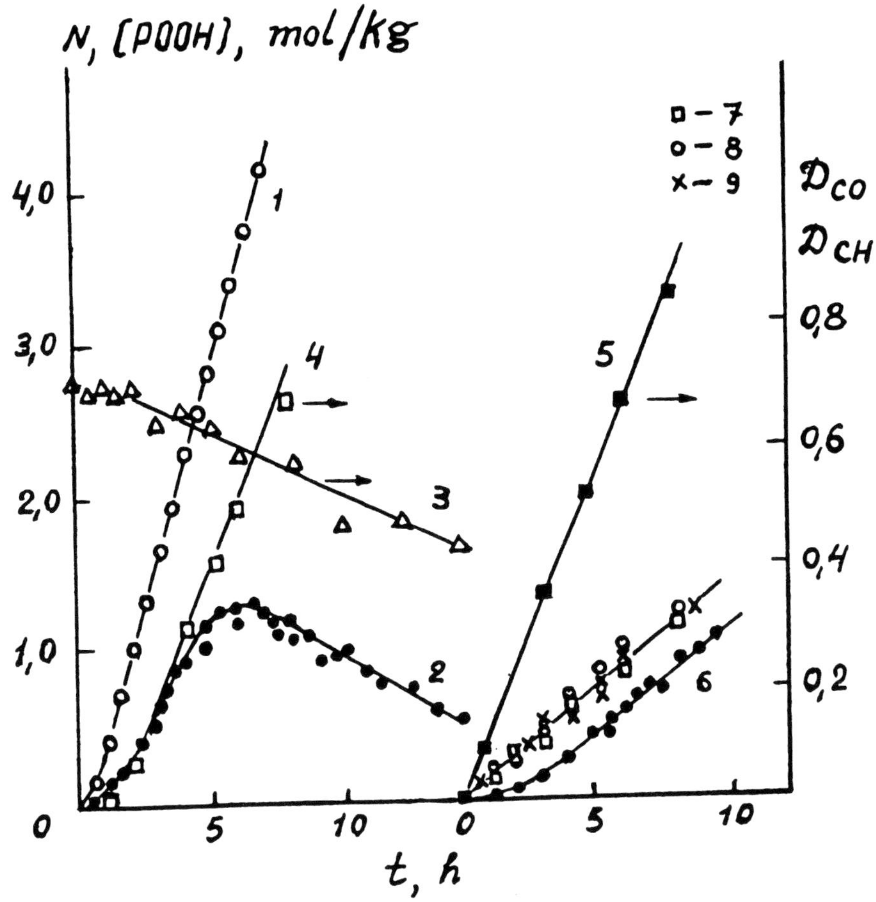

Figure 4. Oxidations curves of melted poly(ethylene oxide) at 90 (1 - 5) and 80 °C (6 - 9): 1 - absorption of oxygen, 2, 6 – 9 - accumulation of hydroperoxide, 3, 4, 5 - change of optical density IR absorption bands of CH - bonds (3) and CO - groups (4, 5). Concentration of the initiator (dibenzoylperoxide) 0 (1 - 4, 6), 0.06 (5), 0.0165 (7), 0.039 (8) and 0.12 mol/kg (9). IR - absorption spectra were registered, using solutions of polymers in chloroform with concentration 10 mg/ml [15, 18].

The fact of independence of stationary oxidation rate from the contents of the doped initiator proofs the presence of such a structural filter ("narrow neck") in a polymeric matrix, limiting transfer of free radicals from those zones, where dissociation of the initiator occurs, into those zones, where the oxidizing reactionary chains proceed.

It is interesting that the rather high contents of BP in PEO melt renders depressive influence on reaction: the increase of its concentration from 0.12 mol/kg up to 0.38 mol/kg reduces a value of V_3 in 1.4 times at 90 °C [19], that is abnormal from the point of view of homogeneous reaction's model.

The effect of the "narrow neck" was precisely shown up in experiments with solid IPP as well. The diagrams published in [21], showing change of total concentration of consumed BP and collecting POOH, are submitted at oxidation IPP films on air at 100 °C are shown on Fig. 5. In this particular case presence of the structural filter for incipient radicals is represented as the obvious fact: then more BP contents in polymer, then more it breaks up for nothing in relation to process of accumulation of POOH. Besides it is seen that the stationary rate of the process does not depend on initial BP concentration and of the current concentration of POOH.

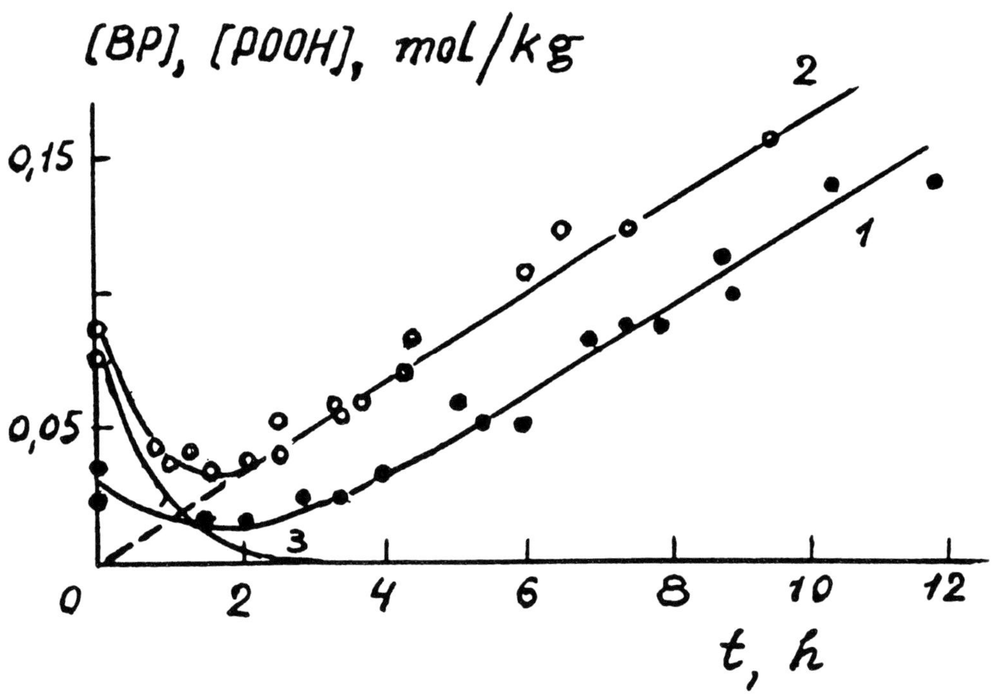

Figure 5. Change of total concentration of dibenzoylperoxide and hydroperoxide (1, 2) and separately dibenzoylperoxide (3) in oxidizing on air films IPP at 100 °C [21].

Let's note one more remarkable result obtained in work [26] at oxidation of low-viscosity melt of atactic polyprolylene (APP) with molecular weight M_n = 4500 (Fig. 6, curves 1, 2) and orientationally stretched films of IPP (Fig. 6, curves 3, 4). The curves 1 and 3 on Fig. 6 related to absorption of oxygen, curves 2 and 4 - to accumulation of POOH (both polymers were oxidized at 120 °C and p = 103 kPa).

The carried out research has shown [26] that the rate of oxidation of the aforenamed objects is increased directly proportionally to concentration of accumulating POOH for a long time. Thus the amount of the absorbed oxygen grows in time according to the exponential law and $V \sim \exp(\Phi t)$. The same dependence is observed at 110 and 130 °C. At the same time solutions of the same polymers in inert trichlorobenzene (in concentration of 0.56 mol/kg calculated on repeating unit) oxidizing according to kinetics of homogeneous reaction, revealing out dependences $V \sim t$ and $V \sim [POOH]^{0.5}$. It turns out that kinetics of oxidation of polymeric objects in a state of increased thermofluctuate dynamics of molecular sponge can differ from kinetics of liquid hydrocarbons and high-molecular weight structural-stationary polyolefins.

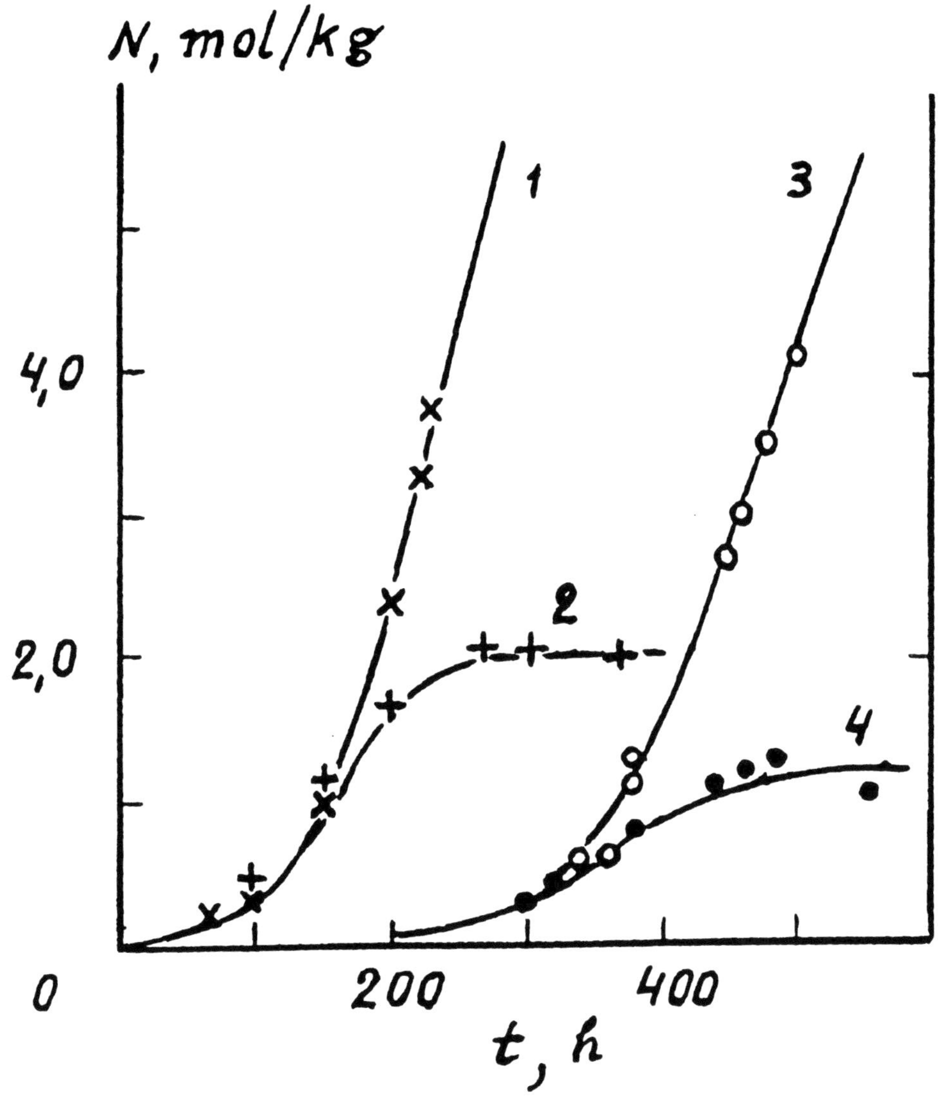

Figure 6. Dependence of quantity of the absorbed oxygen (1, 3) and accumulated hydroperoxide (2, 4) of time of low molecular weight IPP melt oxidation (1, 2) and orientationally stretched films IPP (3, 4) at T = 120 °C, p = 103 kPa [26].

Suming the discussed material we can tell that the oxidation of the same (from the chemical point of view) polymeric reagent is combining (depending on its thermomechanical state) with such a set of qualitatively unequal kinetical laws, which is incompatible with a model of homogeneous reaction. This fact forces to address to new model constructed on the base of nanostructural heterogeneity of a non-crystal polymeric phase and allowing to come with a new logic of theoretical constructions.

4. Model of a Microreactor and Kinetical Modes of Oxidation

In works [15, 16, 18 - 22, 28, 29, 33, 34, 38, 39] matrixes of non-crystalline polymers are considered as a system of molecular-chain spongy micelles, united by a continuous spatial skeleton and filling cell of a skeleton with the size about tens of nanometers (Fig. 7a). The continuous spatial skeleton fixing polymeric chains inside micelles as sponge is composed for a type of a lattice from paracrystalline domains, each of which contains up to ten parallel stacked segments. In a matrix of partially crystalline polymers the spatial skeleton includes also more or less made out fixture from lamellar crystallites (Fig. 7b).

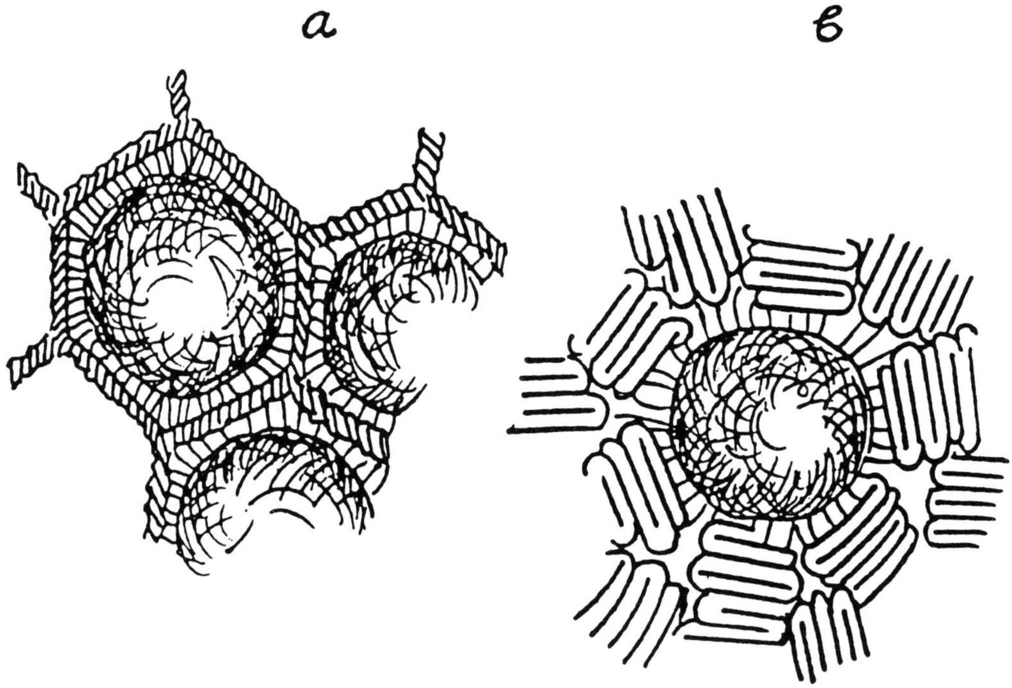

Figure 7. The scheme of micellar-sponge nanostructural reactor for non-crystalline (a) and partially crystalline (b) polymers.

A molecular-sponge micelle, acting as a chemical microreactor, represents the heterophaseous unit, whose chains enter in paracrystalline domains with one end and in a spongy grain with another one. Chains in volume of a grain are pulled together by forces of cohesion, however, lacking a possibility to reach complete collapse they are stabilized as a sponge, filled by rather close micropores (more precise - nanopores) with the sizes from chemical bond up to thickness of a polymeric chain (0.1 - 0.5 nm).

In conditions of balance of elastic forces between a grain and domains of a continuous skeleton the layer from radial oriented chain segments forming walls of rather large (super-) nanopores is formed. As a result of it micelle receives two nanostructural zones: a *s*-zone from supernanopores, having the sizes about a thermodynamic segment (2-3 nm) and *v*-zone - grain with close nanopores.

The chain segments of both zones differ by dynamics of the movements. In a *s*-zone they carry out a role of the mechanical extensions fixing a micellar grain, and *s*-zones itselves can be assimilated to porous adsorbent with a rigid skeleton. The *v*-grain is closer to a more mobile drop of a liquid on thermomechanical properties. This drop (suspended to domains of a continuous spatial skeleton) can in the certain limits make own oscillatory-rotary movements and volumetric pulsations, transferring impulses of movement on composing it more or less flexible polymeric chains.

The spongy character of stacking of flexible chains in micelle's grains gives to a sponge ability to reorganization and capture even of such impurity, whose molecules surpass empty nanopores in the size. Besides, being in essence system of the connected pendula, molecular sponge possess such a property as interference, that is damping or synchronization of the fluctuations, that creates conditions for fluctuational increase of the sizes of *v*-nanopores and, correspondingly, for more or less fast migration of impurity on sponge's volume.

Thin heterophasity of molecular-spongy micelles is reflected on chemistry of chain reactions for some reasons. First, impurities (oxygen, initiators, inhibitors) are allocated non-uniformly between zones, being accumulated mainly in spongy grains. Secondly, differing by the sizes, *s*- and *v*-nanopores provide non-equivalency of a physical state and reactivity at the same reactionary centers belonging to macromolecules or absorbed impurities. Really, in close *v*-micropore the reacting particle will spend the basic time in the constrained conditions either on a nanopore's wall (if it is connected with the macromolecule), or in a pore's volume (if it belongs to an impurity). Thus, the dope molecule repulses off in pore's volume under influence of dispersion electromagnetic forces [28, 29]. At the same time a situation in *s*-nanopore is different: the dope molecule here will spend the basic time in adsorbed state on pore's walls.

Connected with steric and energetic non-equivalency of *s*- and *v*-nanopores a non-equivalency of chemical activity reflects on all intermediates of chain reaction, and the quantitative changes of reactivity of the same reagents reach such values, that a qualitative change of an orientation of their transformations in different zones takes place.

The chain reaction in spongy micelles can proceed in various modes, the most simple for kinetic modeling are stationary-heterophaseous and homophaseous, realized in the purest kind at negligible low rate of structural evolution, that is at practically constant volumes of zones and their properties.

The heterophaseous process corresponds to interface of reactionary chains allocated on both structural zones of spongy micelles. The scheme of heterophaseous process, according to its definition, should include the separate block-diagrams of interzone reactions, as well as schemes of the acts of an interzoned chemical induction in straight and opposite directions. Primary reactions in this process are usually ones in *s*-zones, because the conditions in these zones are most favorable for dissociative initiation of free radicals and they are the donors of free valencies for *v*-zones.

If the chemical reaction is located in one of zones, it is possible to name it formally as a homophaseous one. Such situation occurs, for example, in case of initiation of free radicals in *v*-zones, when *v*-nanopores receive (for whatever reasons) high frequency of fluctuations up to the sizes of *s*-nanopores. In this case POOH groups seized by *s*-nanopores increase its initiating activity and in the certain conditions provide a prevailing way of initiation of chain reaction.

As it was already noted, in absence of obvious structural evolution of high-molecular weight polyolefins the mode of heterophaseous interface of non-uniform reactionary chains is more widespread. Thus the oxidation of polymer without of the dope initiator begins as a result of direct interaction of adsorbed on *s*-nanopores' walls oxygen with CH-groups of polymeric *s*-segments. The basic fraction of the initiated free radicals enters the further reactions and perishes not leaving *s*-nanopores. The process becomes heterophaseous, if the small amount of radicals moves from *s*-zones into volume of spongy grains and induces reactionary *v*-chains in it.

The sequence of the reactionary acts of free radicals' *s*, *v* - translation (characterized by the appropriate scheme) physically expresses a passage of free valencies through structural "narrow neck". A passing ability of this "neck" changes depending on thermomechanical state of a polymeric sample and from presence of impurities capable to promote or, vise versa, to suppress process *s*, *v* - translation of free valencies.

Induced by such transfer in v-zones the secondary reactionary chains differ from primary by qualitatively different transformations therefore, in particular, that it contains more oxygen and the speed of polymeric chains' cleavage with participation of alkyl macroradicals is lowered. The reactionary oxididative *v*-chains have very low termination rate in comparison with reactionary *s*-chains owing to equivalently low rate of *s*, *v* -translation of free radicals, and it promotes their large length. In such circumstances, amount and the structure of products of oxidation is determined mainly by transformations of polymeric parts of v-zones making basic weight of molecular-sponge micelles.

As to a reverse chemical induction, in oxidizing process it is connected to transfer of hydroperoxide groupings from *v*-zones in *s*-zones. The hydroperoxide formed in close *v*-nanopores differs by a very low rate of dissociation on radicals owing to adverse for this purpose steric and energetic conditions. However during the accumulation in sponge hydroperoxide occupies stationary *s*-zones as well with their supernanopores favorable for dissociation, translating process of oxidation from the induction period in a stage of autoacceleration.

5. Reactionary Model of an Oxidation Induction Period

According to aforetold, the general scheme of the initial period of autooxidation of polyolefins includes two blocks, for *s*- and *v*-zones respectively, and also scheme of an interzoned chemical induction correlated for interface of non-uniform reactionary chains.

In the absence of the dope initiator the initiation of free radicals is carried out in *s*-nanopores by interaction of oxygen with polymer's CH-groups. An exchange of molecules of oxygen between zones

$$O_{2,v} \leftrightarrow O_{2,s}$$

proceeds quickly, not experiencing of any diffusion difficulties (the indexes *s* and *v* designate belonging of reagent to the appropriate zone). Thus the basic amount of oxygen is accumulated in v-zones, where its concentration practically does not differ from concentration in a non-crystalline phase: $[O_{2,v}] \approx [O_2]$.

The scheme of a primary reactionary chain can be written in the following general view:

$$PH_s + O_{2,s} \xrightarrow{k'_{0s}} \delta_1 P^{\bullet}{}_s \text{ (radicals' initiation)}$$

$$P^{\bullet}{}_s \xrightarrow{k_d} R^{\bullet}{}_s + R^{=} \text{ (a cleave of polymer chain)}$$

$$R^{\bullet}{}_s \xrightarrow{k_{1s}} R^{\bullet}{}_{1s} \text{ (isomerization)}$$

$$R^{\bullet}{}_s + R^{\bullet}{}_{1s} \rightarrow \text{the products of recombination and disproportionation}$$

Designations here used are: PH – repeating unit of a chain with reactionary hydrogen atoms, $P_s^{\bullet}$ - macroradical with free valency on medial repeating units of macromolecules (its structure for polyethylene looks ~$CH_2C^{\bullet}H$~, for polypropylene - ~$CH_2C^{\bullet}(CH_3)$~ and ~$CH^{\bullet}CH(CH_3)$~), δ_1 - coefficient of an yield of free radicals, $R_s^{\bullet}$ and $R^{=}$ - a terminal radical and unsaturated group, $R_{1s}^{\bullet}$ - isomerization product of terminal $R_s^{\bullet}$ (structure of radicals $R_s^{\bullet}$ and $R_{1s}^{\bullet}$, related to PP, is considered in work [21]).

The scheme of interzoned transfer of free valency includes the acts of an yield of a radical $R_s^{\bullet}$ from structure of a supermolecular skeleton in a volume of a s-nanopore (in state of $R_m^{\bullet}$) and subsequent oxidation:

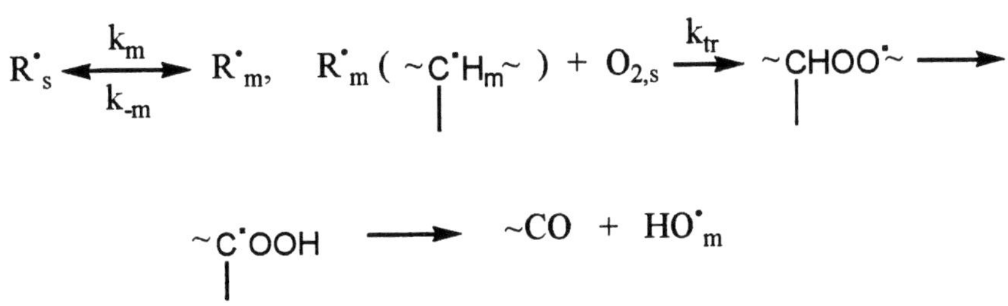

here vertical line at atom C designates presence of the attached CH_3 group on PP chains or atom H on PE chains.

As it was already noted the rate of s, v-transfer is negligibly low in comparison with rate of transformations of s-radicals. It is determined by radicals' thermomechanical ability $R_s^{\bullet}$ to oscillate with certain frequency between a rigid lattice composed from s-segments and volume of s-nanopore and, thus to occupy a convenient state $R_m^{\bullet}$ for interaction with oxygen. Formed by such way termicnal peroxyl then undergoes α-isomerization [40] in a hydroperoxyalkyl, decomposing with emission of light free hydroxyl.

Migrating into a v-zone, free hydroxyl induces chain reaction in it, including hydroperoxide and epoxide branches:

$$HO^{\bullet}_{m} \longrightarrow HO^{\bullet},\ HO^{\bullet} + PH \xrightarrow{k'_1} H_2O + P^{\bullet},\ P^{\bullet} + O_2 \xrightarrow{k_1} POO^{\bullet},$$

$$POO^{\bullet} + PH \xrightarrow{k'_2} POOH_v + P^{\bullet},$$

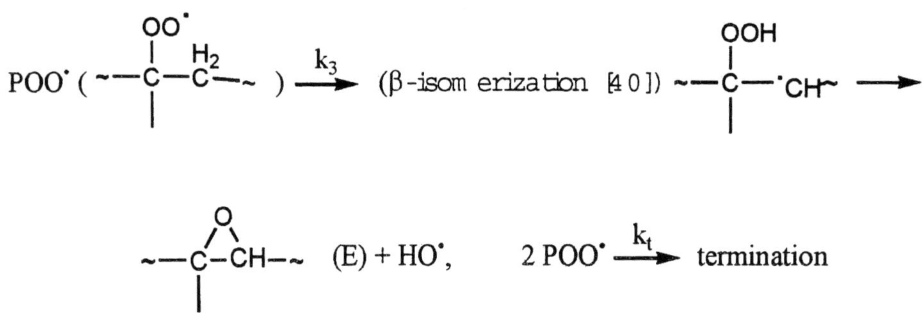

Epoxides (E), having high reactivity to free radicals, enter the further transformations with formation of carbonyl compounds (appropriate reactions will be discussed below).

Applying a condition of stationary concentration for free radicals at consideration of the s-scheme, we shall not take into account losses of s-radicals on interzoned transfer, but we shall take into account the fact of transfer as a stage of initiation at consideration of the reactionary v-scheme. According to this we shall express concentration of macroradicals, reflecting an interzoned chemical induction

$$[R^{\bullet}{}_s]_1 = k''_{0s}[O_{2,s}]/2k_{1s}$$

rate of s, v – transfer

$$V_{tr,1} = k_{tr}k_m k''_{0s}[O_{2,s}]^2/2k_{1s}(k_{-m} + k_{tr}[O_{2,s}])$$

and concentration of peroxiles in v-zones on a 1st stage of oxidation

$$[POO^{\bullet}]_1 = [k_{tr}k_m k''_{0s}/4k_t k_{1s}(k_{-m} + k_{tr}[O_{2,s}])]^{0.5}[O_{2,s}] \qquad (2)$$

where $k''_{0s} = \delta_1 k'_{0s}[PH_s]$ (the subscript index 1 designates the first stage of oxidation process).

Using the equation (2), we shall express rates of oxygen absorption (V_1), accumulation of hydroperoxide ($V_{g,}$ 1) and discharge of water (V_{aq}, 1):

$$V_1 = -d[O_2]/dt = k_1[P^\bullet][O_2] = (k'_2[PH] + k_3)[POO^\bullet]_1 = (k_2 + k_3)ap/(b+p)^{0.5} \quad (3)$$

$$V_{g,1} = dg/dt = k'_2[PH][POO^\bullet]_1 = k_2 ap/(b+p)^{0.5}$$

$$V_{aq,1} = k'_1[HO^\bullet][PH] = k_3[POO^\bullet]_1 = k_3 ap/(b+p)^{0.5}$$

agrees with experiment [15, 20 - 22]; here $a = (k_m k''_{0s}\gamma_s / 4k_t k_{1s})^{0.5}$, $b = k_{-m}/k_{tr}\gamma_s$, $k_2 = k'_2[PH]$, $\gamma_s = [O_{2,s}]/p$.

Let's emphasize that in opposite to homogeneous model in the heterophaseous mechanism of oxidation water is formed mainly not by of disintegration POOH, but by chain reaction with participation of free hydroxyls.

The transformations connected to oxidation of intermediate epoxides, result to additional absorption of oxygen, formation of POOH and water, but do not affect on the equation (2). Therefore, expressions written above for macroscopic rates remain the same.

Epoxides receive rather low stationary concentration owing to active interaction with free radicals:

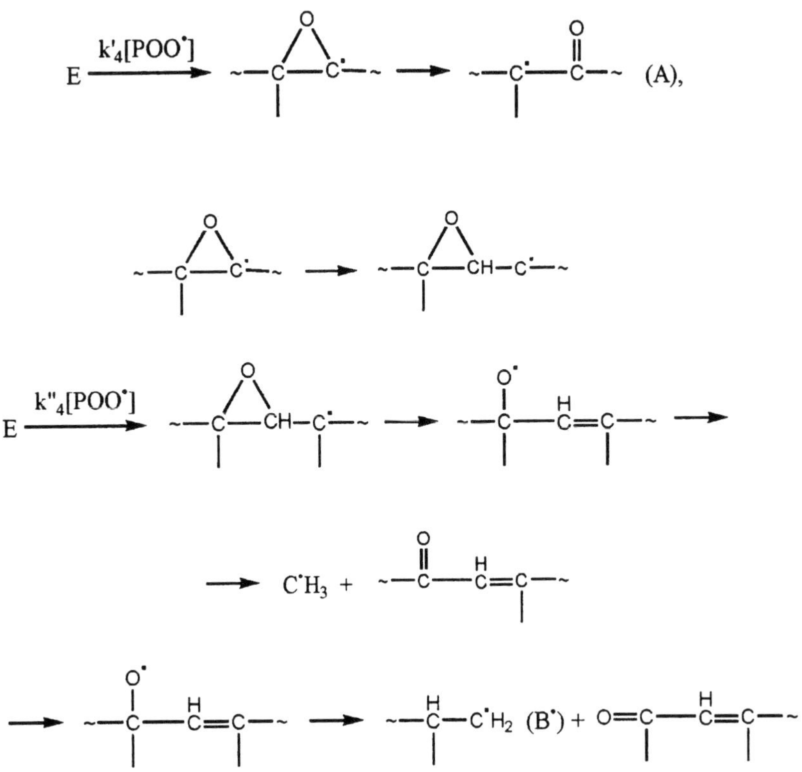

The latter provides macromolecules' destruction and formation of carbonyl groups (CO). The products of epoxides' oxidation are found out with some delay compare to POOH [15, 21], however rate of their formation is significant enough already in the induction period, and then their rate is quickly increases and is stabilizes for a long time [41 - 43]. Composition of formed CO - compounds is rather various (polymeric and volatile aldehydes, ketones, and acids are among them) and can be detailed by the following reactions:

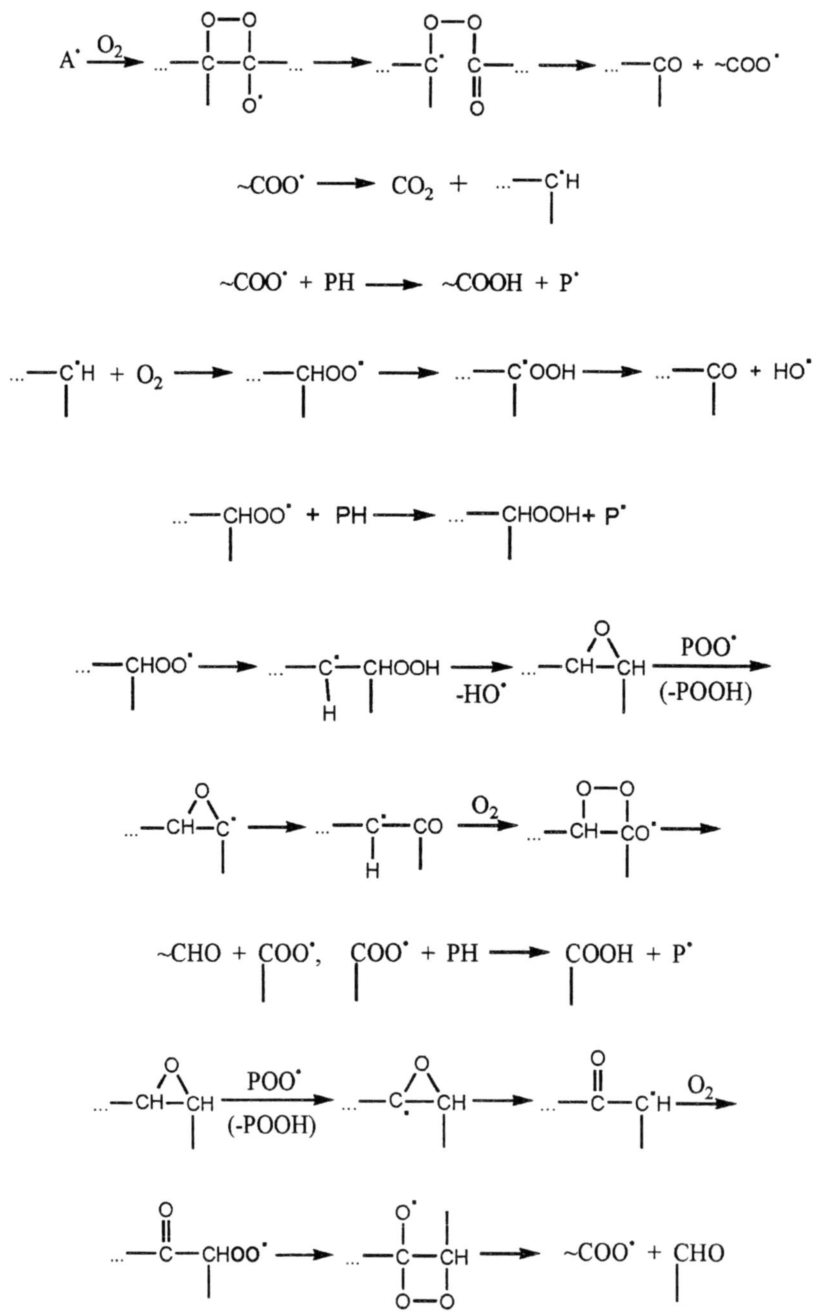

Isomerization of the terminal radical CH$^{\bullet}$ provides moving of free valency along of macromolecule and formation of ketone groups on increased distance from the end of a chain:

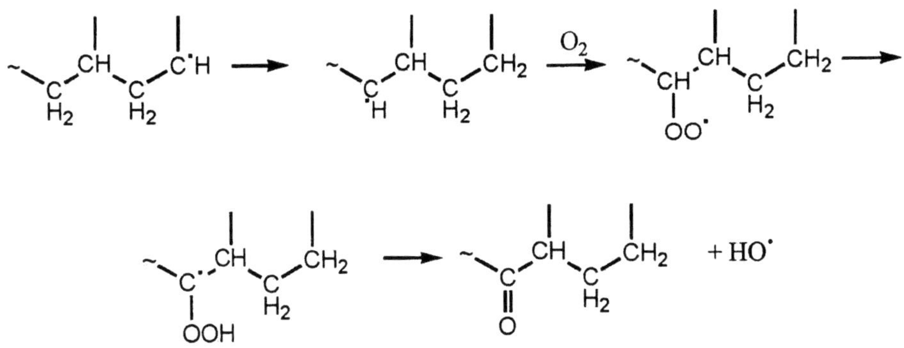

The further oxidation of these ketones results to formation of larger fragments with aldehyde, keto- and acidic groups:

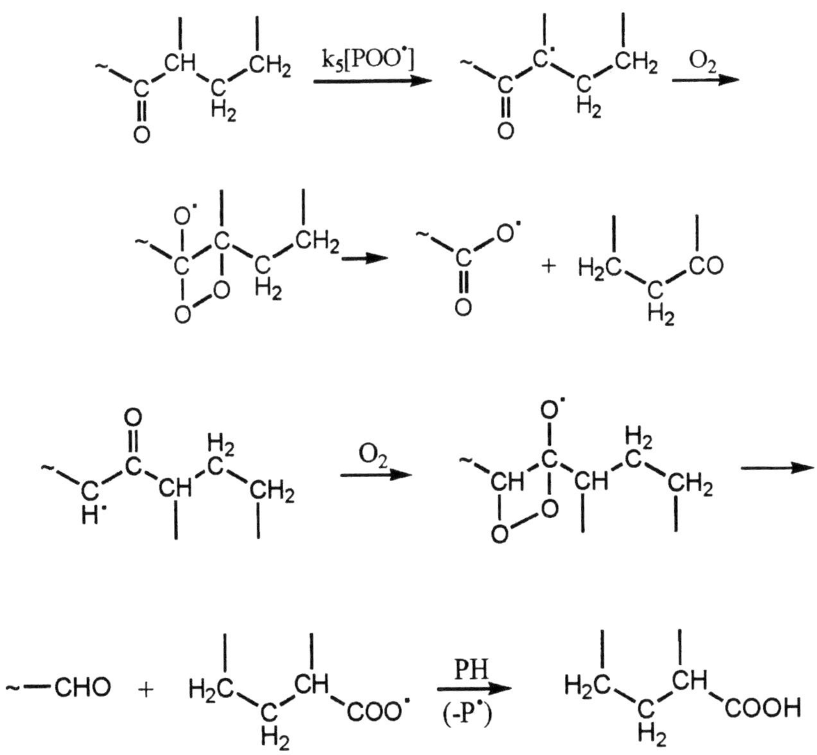

The written reactions are proved by presence of such compounds as methylethylketone, methylpropylketone, butyraldehyde, pentanone, hexanone etc. among products of polyolefines' oxidation [41 - 43].

Transformations of a terminal radical $B^{\bullet}$ can be presented by the following scheme:

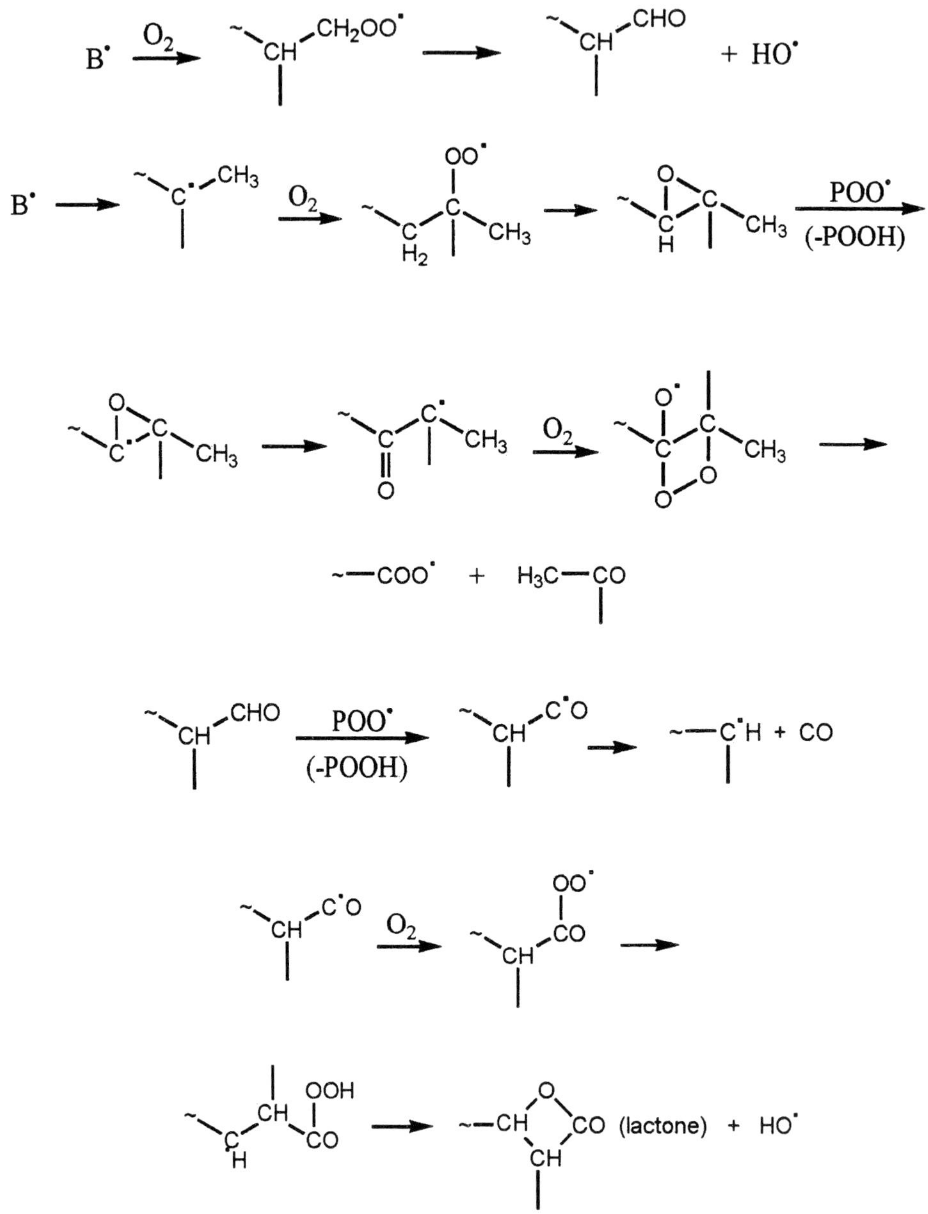

The made incomplete record of reactions is sufficient for demonstration of ways of formation of final products of polyolefines' oxidation and that fact, that transformations of radicals $A^{\bullet}$ and $B^{\bullet}$ (as well as CO – compounds) do not influence on stationary concentration of peroxiles (equation (2)). CO - compounds continuously increase their concentration being oxidized and arising again in chain process.

Thus, a condition of stationary concentration of compound E can be written:

$$dE/dt = k_3[POO^{\bullet}]_1 - k_4[E][POO^{\bullet}]_1 = 0, \qquad (4)$$

where $k_4 = (k_4^{'} + k_4^{''})$, $[E] = k_3 / k_4 = const$ and to express total rate of accumulation of CO - compounds, formed in epoxy branches

$$V_{CO,1} = d[CO]/dt = (\alpha_1 + \alpha_2)k_4[E][POO^{\bullet}]_1 = \alpha k_3[POO^{\bullet}]_1$$

where α_1, α_2 - yield of polymeric and volatile compounds respectively, α - their sum. Thus, the rate $V_{CO,1}$ receives the same dependence of oxygene's pressure, as rates V_1 and $V_{g,1}$, after achievement of a condition of equation (4) and the contents of CO - groups is increased proportionally to time

$$[CO]_1 = V_{CO,1}t \tag{5}$$

The same relates to process of accumulation of concentration (n) of cleaved polymeric chains in ν-zones of spongy micelles. The cleave rate develops of some fractions from the rates which have been written with the constants k'$_4$, k''$_4$:

$$V_{n,1} = dn/dt = (\nu_1 k_4^{'} + \nu_2 k_4^{''}[E][POO^{\bullet}]_1 = const[POO^{\bullet}]_1 \tag{6}$$

where ν_1, ν_2 - coefficients.

Following from the scheme of the induction period the direct proportionality between amount of oxidation's products and amount of the absorbed oxygen remains the same for the certain time after the induction period as well. This time is significantly longer for processes of accumulation of CO - compounds and destruction of macromolecules in comparison with process of POOH accumulation [41 - 43].

It is necessary to emphasize that the equation (5) characterizes total process of CO – compound's formation, accompanying by splitting of low-molecular weight fragments off. Such acts of destruction poorly affect on a polymeric solution's viscosity in comparison with primary destructive acts in epoxy branches, which rate is described by the equation (6).

6. Model of Reaction Proceeding Under Control of POOH

The induction period of polyolefin autooxidation comes usually to an end after absorption of several hundredth mol/kg of oxygen owing to settling of s-zones of a sponge matrix by hydroperoxide, coming from v-zones. The process of settling is provided with chemical relay race of free valency, migration of OOH-groups along macromolecules and by fluctuational acts of rearrangement of molecular sponge [15, 20, 21, 33].

Designating total frequency of processes of v, s- transfers of OOH-groups through an effective constant k_{vs} and taking into account that for activization of these processes the acts of sponge fluctuation with opening of tight v-micropores are necessary, we shall write the scheme of v, s- transfer of hydroperoxide,

$$POOH_v \underset{k_{-f}}{\overset{k_f}{\leftrightarrow}} POOH_f \xrightarrow{k_{vs}} POOH_s$$

where the index f relates to fluctuationally exited v-nanopore with the seized OOH-group.

The OOH groups receive an opportunity to dissociate on free radicals getting in nanopores of s-zones

$$POOH_s \xrightarrow{k_{0s}} (HO^\bullet + PO^\bullet)_s \xrightarrow{(PH_s)} \delta P_s^\bullet$$

and enter in bimolecular reaction in process upon increase of its concentration

$$POOH_s + POOH_f \xrightarrow{k^f} 2POH + O_2$$

without formation of free radicals. The given scheme of reverse chemical interzoned connection allows to express stationary concentration of OOH-groups in their fluctuationally active and in s-conditions:

$$dg_f / dt = k_f g_v - k_{-f} g_f - k_{vs} g_f - k^f g_s g_f = 0$$

Here it is possible to neglect by two last components in comparison with high rates of fluctuation deactivation, and to use deducted expression $g_f = k_f g_v / k_{-f}$ in the equation of rate of s-hydroperoxide reactions.

$$dg_s / dt = k_{vs} g_f - k^f g_s g_f - k_{0s} g_s = 0$$

The found thus concentration

$$g_s = k_{vs} k_f g_v / (k_{-f} k_{0s} + k^f k_f g_v) \tag{7}$$

allows to express total rate of radicals' origin in s-zones

$$V_{is} = k_0^{''}[O_{2,s}] + \delta k_{0s} g_s \approx \delta k_{0s} g_s.$$

The contribution of direct reaction of CH - bonds with oxygen quickly becomes negligibly small and earlier received expression $V_{tr,1}$ is replaced on $V_{tr,2}$:

$$V_{tr,2} = k_{tr} k_m \delta k_{0s} g_s [O_{2,s}] / 2k_{1s}(k_{-m} + k_{tr}[O_{2,s}]).$$

Expression of peroxiles' concentration simultaneously changes also on:

$$[POO^{\bullet}] = [k_{tr}k_m\delta k_{0s}g_s[O_{2,s}]/4k_tk_{1s}(k_{-m} + k_{tr}[O_{2,s}])]^{0.5}. \tag{8}$$

On some initial step of the accelerated stage of autooxidation the final products of reaction are collected synchronously with a hydroperoxide. The latest, not having an opportunity to dissociate into v-zones of a spongy matrix, accumulates during some time practically not reacting with other products of oxidation by virtue of their low concentration. At this time rate of accumulation of hydroperoxide is characterized by the formula:

$$V_{g,2} = k_2[POO^{\bullet}] = A(\delta k_{0,s}g_s)^{0.5}$$

(where $A = k_2[k_{tr}k_m[O_{2,s}]/4k_tk_{1,s}(k_{-m} + k_{tr}[O_{2,s}])]^{0.5}$), or (taking into account the equation (7) and ratio $g_v \approx g$ {the major quantity of OOH-groups of a non-crystalline polymeric phase is concentrated in v-zones}):

$$V_{g,2} = A[\delta k_{0s}k_{vs}/k^f(1 + k_{0s}k_{-f}/k^fk_fg_v)]^{0.5} = V_{g,3}/(1 + c/g)^{0.5} \tag{9}$$

where $V_{g,3} = A(\delta k_{0s}k_{vs}/k^f)^{0.5}$, $c = k_{0s}k_{-f}/k^fk_f$.

The expressions of rates of other macroscopic processes look like:

$$V_{co} = \alpha k_4[E][POO^{\bullet}] = V_{co,3}/(1 + c/g)^{0.5} = \gamma V_{g,3}/(1 + c/g)^{0.5} \tag{10}$$

$$V_{aq} = k_1'[HO^{\bullet}][PH] = \beta k_3[POO^{\bullet}] = V_{aq,3}/(1 + c/g)^{0.5}$$

$$V = V_{co} + V_{g,2} = V_3/(1 + c/g)^{0.5} = \varphi^{-1}V_{g,3}/(1 + c/g)^{0.5} \tag{11}$$

$$V_n = (v_1k_4' + v_2k_4'')[E]]POO^{\bullet}] = V_{n,3}/(1 + c/g)^{0.5} \tag{12}$$

where $\gamma = \alpha k_3/k_2$, β - yield of water, φ - yield of hydroperoxide, $V_{g,3}$, $V_{co,3} = \gamma V_{g,3}$, $V_{aq,3}$, $V_3 = V_{g,3}/\varphi$, $V_{n,3}$ - maximal rates on the third macroscopic stage of autooxidation, reached when $c/g < 1$. The latter transforms equation (7) to

$$g_s = k_{vs}/k^f = const \tag{13}$$

The equation (13) reveals, that the autoaccelerated stage of oxidation should sooner or later end, as the s-zone is structural border of micellar reactor and, stopping a flux of onflowing OOH-groups, actuates bimolecular reaction which does not generate free radicals.

The oxidation again proceeds in a mode of zero order on hydroperoxide under control of its non-radical bimolecular reaction, proceeding in s-zones. Now, realizing function of the initiator, hydroperoxide simultaneously carries out a role of the specific buffer, supporting constant rate of initiation of reactionary v-chains and stabilizing concentration of peroxides (equation (8)) on a value independent of time.

The theoretical equations (9 - 11, 13) were experimentally confirmed in works [15, 18, 20 - 22], for the first time given definition for third macrokinetical stage of autooxidation with zero order on hydroperoxide.

The maximal rate of oxygen's absorption V_3 depends on oxygen pressure at a stage of the zero order, aspiring to a limit at high values of p: $V_3 \sim [p/(b+p)]^{0.5}$

In opposite to V_3, the rate V_1 (in the induction period) does not saturate when p increases, turning in $V_1 \sim p^{0.5}$ dependence according to the equation (3). These distinctions did not receive explanations in homogeneous model of process, whereas heterophaseous model assumes that the oxygen initiates oxidation chains twice: once at origin of s-radicals, second time in the acts of s,v-transfer of free valency. At a stationary stage of process the oxygen is not a main initiator anymore, but remains to be a promoter of the acts of radicals' passage through structural "narrow neck".

7. Mechanism of POOH Decomposition by Products of Oxidation

As it was already noted earlier, the synchronism of processes of oxygen's absorption and hydroperoxide's accumulation is invalid from some moment owing to interaction of POOH with the reaction's products. (We shall note, that the decomposition of POOH in oxydized polyolefin in absence of oxygen proceeds by chain reaction [16, 34], in opposite to unchain disintegration of POOH in oxydizing polymer).

It is known, that hydroperoxides are decomposed by carboxylic acids in both polymers and in liquid solutions [3, 44, 45]. Research, carried out in work [45], of mixes of cumylhydroperoxide with acetophenone, dimethylphenylcarbinol and bensoic acid has shown that only acid considerably accelerates disintegration of hydroperoxide with simultaneous formation of free radicals. In oxydizing polyolefins acids are one of the major products of oxidation and are characterized by wide distribution on molecular weights. It may be expected, that acids with long molecular chains built - in in a sponge will not render essential influence on POOH, as having not freedom of moving on a matrix. Fragmented labile acids formed at oxidizing cleavage of end keto – groups most likely should act as a catalyst of disintegration.

Following from heterophaseous scheme the equation of formation rate of end ketones

$$d[CO]_t/dt = \alpha_t k_4[E][POO^\bullet] = \alpha_t k_3[POO^\bullet] = \gamma_t V_{g,3}/(1+c/g)^{0.5}$$

(where α_t - yield (CO), $\gamma_t = \alpha_t k_3/k_2$) can be simplified taking into account that the accelerated stage on oxygen's absorption curves (and accumulation of CO - groups) is much shorter than the period of time with constant maximal rate (Fig. 2) [15, 21, 41 - 43, 46, 47]:

$$d[CO]_t / dt = \gamma_t V_{g,3} = const\,, \qquad [CO]_t = \gamma_t V_{g,3} t\,.$$

According to this the equations for formation rate of labile acids (AH)

$$d[AH]/dt = k_5[CO]_t[POO^\bullet] = k_5\gamma_t V_{g,3} t$$

and for their quantity

$$[AH] = Bt^2, \tag{14}$$

where $B = 0.5k_5\gamma_t V_{g,3} = const$. Square-law on time equation (14) explains the accelerated accumulation of an acidic blend on the latest stages of oxidation of polyolefon, despite of significant decrease of POOH concentration after a point $[POOH]_{max}$ [48, 49].

Addressing to the mechanism of acid-catalyzed decomposition of hydroperoxide it should be taken into account that the expenditure of POOH, accelerated in time, does not affect on constant rate V_3 of a stage of the zero order of autooxidation, that testifies about neglegebly low yield of free radicals in catalytic reaction.

The adequate reactionary scheme of catalysis should include reaction of condensation of POOH with an acid migrating in sponge

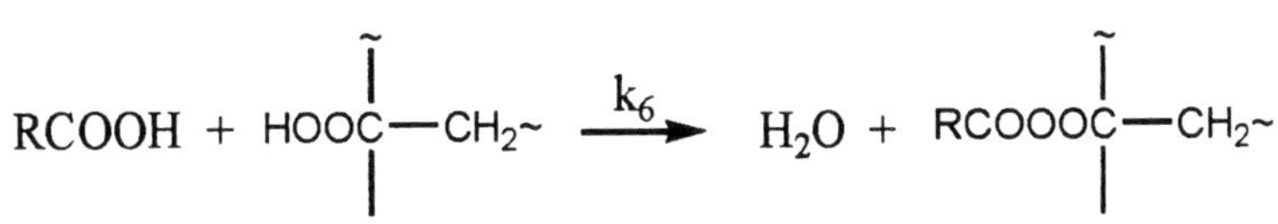

Formed thus peroxyester has to dissociate at a stage of fluctuational expansion of occupied v-nanopore towards to an inner space of s-nanopore

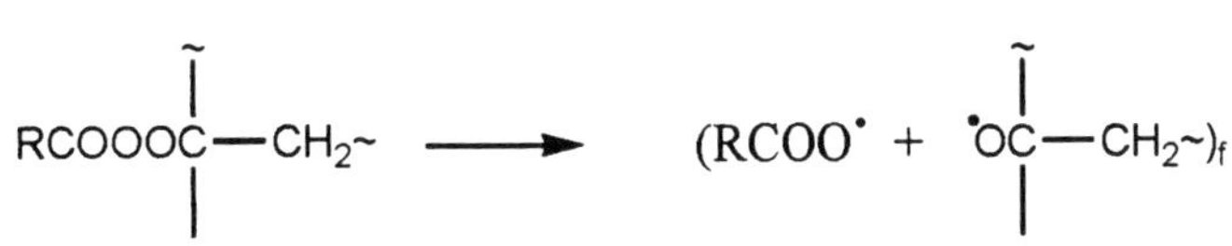

while the generated radicals have to disproprotionate with regeneration of an acid, not having time to leave from increased in the sizes "rigid cage":

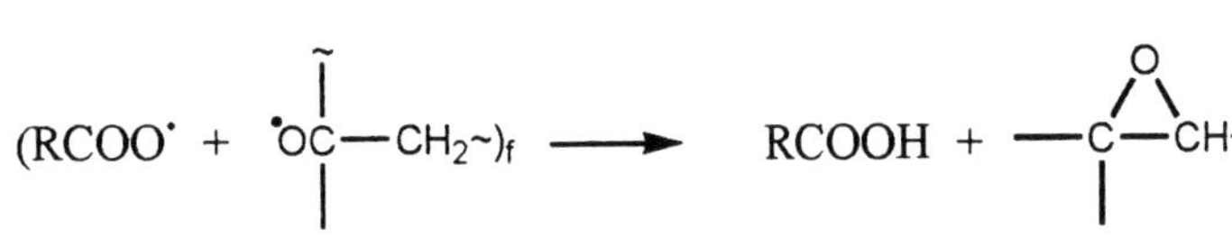

The given scheme of catalysis is equivalent to intracell reaction, at which the radicals do not leave for inner space of reacting material [16].

According to aforetold more complete kinetic equation for rate of hydroperoxide's accumulation is deducted

$$dg/dt = V_{g,3}/(1+c/g)^{0.5} - k_6[AH]g$$

and when expression (14) is taken in account:

$$dg/dt = V_{g,3}/(1+c/g)^{0.5} - Kgt^2 \quad (15)$$

here $K = 0.5k_6k_5\gamma_tV_{g,3}$, and the value $V_{g,3}$ at saturating pressure of oxygen becomes equal

$$V_{g,3} = k_2(\delta k_{0s}k_{vs}k_m/4k^f k_t k_{1s})^{0.5}. \quad (16)$$

In the developed reaction the equation of rate of oxygen's absorption remains former appearance

$$V = k_1[P^\bullet][O_2] = (k_2+\eta k_3)[POO^\bullet] = V_3/(1+c/g)^{0.5} = \varphi^{-1}V_{g,3}(1+c/g)^{-0.5} \quad (11)$$

where η - ratio of acts of oxygen's absorption in epoxy branches of oxidation, φ - yield of POOH at absorption of 1 mol/kg of oxygen.

Calculated verification of kinetic differential equations (11) and (15) carried out with use of the published experimental data on oxidation of melt of APP [50], LDPE [9], copolymers of ethylene with propylene (CEP) [51], solid films IPP-1 and IPP-2, distinguished by the size of spherolites [24, 31], and also results of polypropylene's oxidation of domestic manufacturers (IPP-3), given by L.N.Guseva. The theoretical curves (they are given on Fig.2, 8 - 11) calculated by a Runge-Cutta method in the simplified mode, integrating separately equation (11) and equation (15) [52]. The given simplification is accepted instead of the solution of two equations system, limits an interval of calculations in the field of the high degree of oxidation by an inequality **c**/g < 1, that it is quite enough for a soluble task.

The necessary values of V_3 and $V_{g,3}$ estimated on tangent line, carried out to ascending parts of curves Fig. 2, 8 - 11, the factors **c** and K searched from mathematical properties of the equations (11), (15). Owing to much scatter of experimental points, and also values of the induction periods (τ) and adequate to the moments τ of concentrations of accumulated hydroperoxide (g_1) all listed characteristics were specified, conducting series of calculations down to the best conformity of calculated curves with experimental points.

The diagrams of dependence of POOH concentration in oxidizing in air APP at temperatures 90, 100, 110 and 120 °C [50], are given on Fig. 8 (curves 1 - 4). Kinetic characteristics corresponding to the curves are given in the following numbers: τ = 33, 10, 2 and 0.1 h; $V_{g,3}$ = 0.0129, 0.0255, 0.0400, and 0.0700 mol/ (kg*h); K = $6.3*10^{-6}$, $5.6*10^{-5}$, $3.0*10^{-4}$ and $2.0*10^{-3}$ h^{-3}, in all cases an equality **c** = g_1 = 0.02 mol/l is carried out. The formulas, generalized on all temperature interval, of the specified values (given to a minute time scale) look like:

τ (min) = $5.25*10^{-19}$ exp (E_τ/RT), E_τ = 150 kJ/mol;

$V_{g,3}$ (mol/l*min) = $2.1*10^{6}$ exp (-E_g/RT), E_g = 69.5 kJ/mol;
K (min^{-3}) = $3.9*10^{22}$ exp (-E_k/RT), E_k = 230 kJ/mol

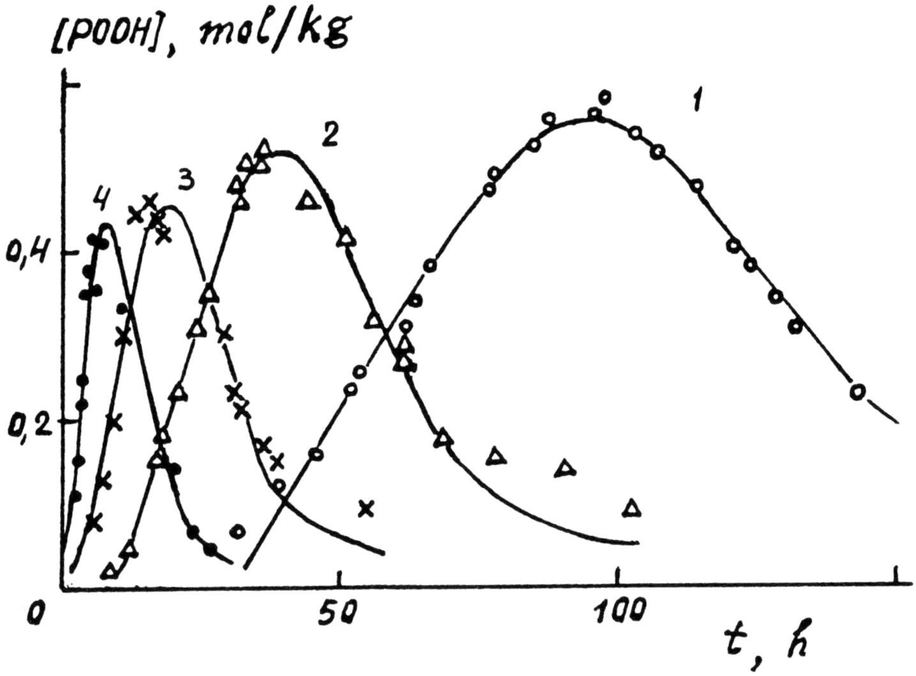

Figure 8. Calculated curves of hydroperoxide's accumulation during oxidation of APP on air at temperature 90 (1), 100 (2), 110 (3) and 120 °C (4) [50].

The calculated curve of oxygen's absorption (1, 3) and POOH accumulation (2, 4) in LDPE melt, oxidizing at pressure of oxygen 100 kPa are given on Fig. 9. Values of τ = 160 min, V_3 = 0.006 mol/kg*min, $V_{g,3}$ = $1.05*10^{-3}$ mol/kg*min, K = $1.7*10^{-8}$ min^{-3} correspond to curves 1 and 2 (130 °C), and values τ = 12 mines, V_3 = 0.076 mol/kg*min, $V_{g,3}$ = 0.0145 mol/kg*min, K = $1.6*10^{-5}$ min^{-3} correspond to curves 3 and 4 (160 °C). The curves, not given by us, for LDPE, obtained in work [9] at 150 °C (*p* = 100 kPa), well correlated to experiment at values of τ = 35 min, V_3 = 0.036 mol/kg*min, $V_{g,3}$ = 0.005 mol/kg*min, K = $1.0*10^{-6}$ min^{-3} (for LDPE the equality **c** = g_1 = 0.02 mol/kg takes place as well). The appropriate generalized formulas are:

τ (min) = $4.34*10^{-15}$ exp (E_τ/RT), E_τ = 128.4 kJ/mol;

V_3 (mol/l*min) = $8.86*10^{13}$ exp (-E/RT);

$V_{g,3}$ (mol/l*min) = $1.48*10^{13}$ exp (-E_g/RT),

$E = E_g = 124.4$ kJ/mol,

$K\ (\text{min}^{-3}) = 1.86 * 10^{31} \exp(-E_k/RT)$, $E_k = 300$ kJ/mol

an average value of an hydroperoxide yield is $\varphi = 0.167$.

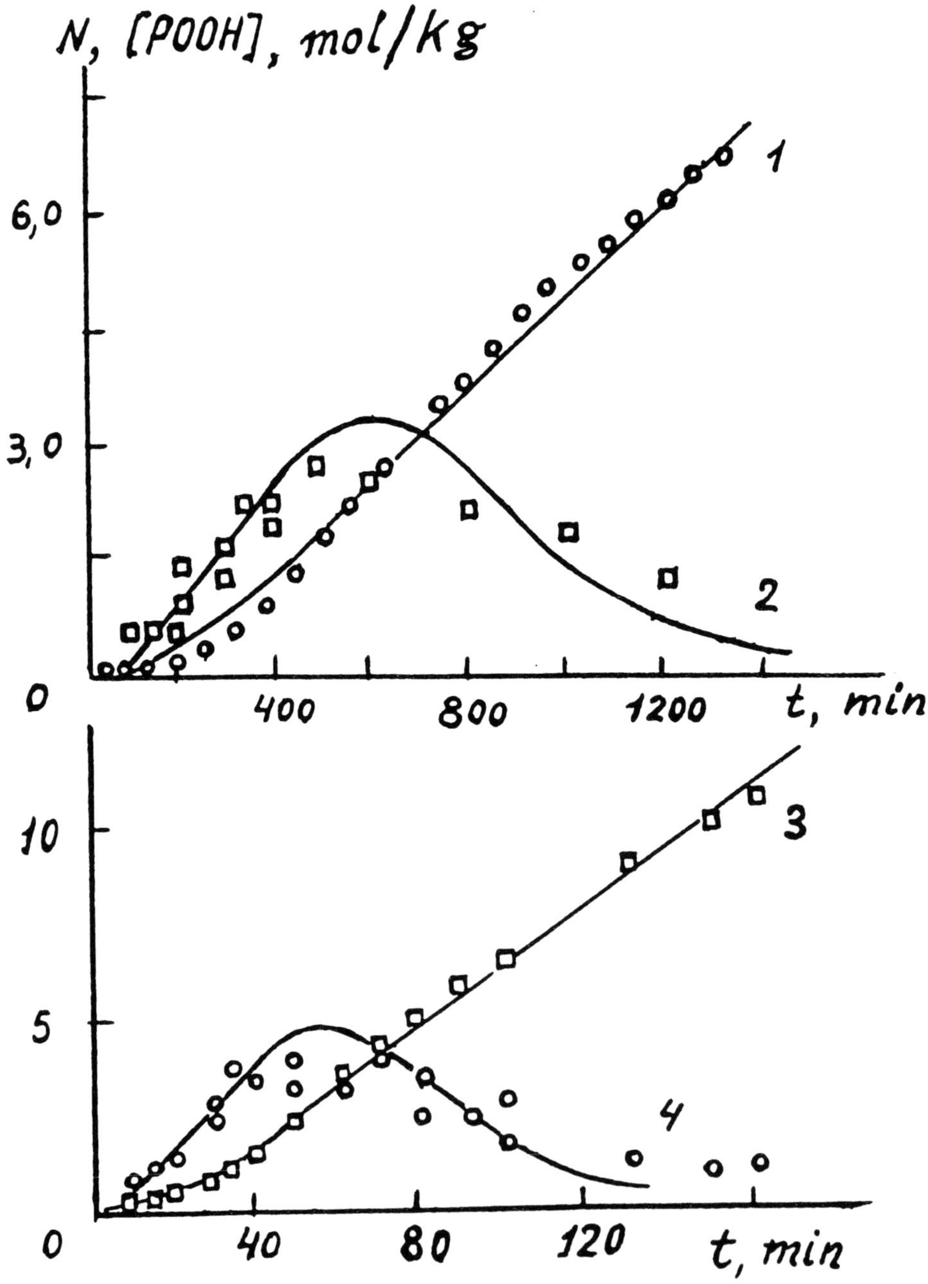

Figure 9. Calculated curves of oxygen's absorption (1, 3) and hydroperoxide's accumulation (2, 4) during oxidation of LDPE melt at temperature 130 (1, 2) and 160 °C (3, 4), $p = 100$ kPa [9].

The calculated curves 1, 3 of oxygen's absorption and curves 2, 4 of POOH accumulation in copolymers CEP-1 and CEP-2, investigated in work [51] and containing in its macromolecule structure accordingly 1.5 and 24 weight percent of polypropylene chains (130 °C, p = 40 kPa) are submitted on Fig. 10. In this case calculations are well correlated to experiment at mentioned below kinetic characteristics.

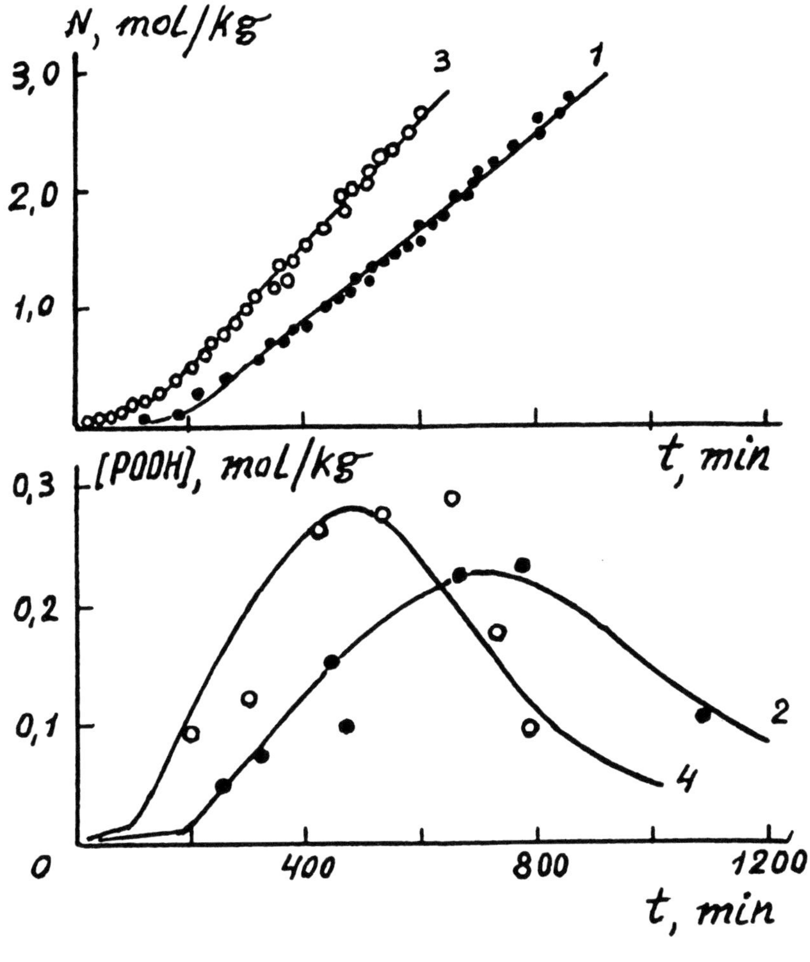

Figure 10. Calculated curves of oxygen absorption (1, 3) and hydroperoxide's accumulation (2, 4) in melt of CEP - 1 (1, 2) and CEP - 2 (3, 4) at 130 °C and p = 40 kPa [51].

CEP-1 (Fig. 10, curves 1, 2) **c** = g_1 = 0.01 mol/kg, τ = 180 min, V_3 = 0.004 mol/kg*min, $V_{g,3} = 6.2*10^{-4}$ mol/kg*min, $K = 1.0*10^{-8}$ min^{-3}, φ = 0.15.

CEP-2 (Fig. 10, curves 3, 4) **c** = g_1 = 0.02 mol/kg, τ = 100 min, V_3 = 0.005 mol/kg*min, $V_{g,3} = 1.1*10^{-3}$ mol/kg*min, $K = 2.8*10^{-8}$ min^{-3}, φ = 0.22.

The similar calculated diagrams for samples IPP-1 (with fine spherolites) and IPP-2 (with large spherolites), oxidized at 130 °C and p = 40 kPa are given on Fig. 2 (accordingly curves 1, 5 and 2, 6; just a reminder that the oxygen curves are given by the authors without experimental points). The calculations have shown presence of satisfactory conformity between the theory and experiment at the below-mentioned characteristics.

IPP-1: **c** = 0.025 mol/kg, g_1 = 0.01 mol/kg, τ = 10 min, V_3 = 0.365 mol/kg*min, $V_{g,3}$ = 0.005 mol/kg*min, K = 3.0*10^{-6} min^{-3}, φ = 0.14.

IPP-2: **c** = 0.025 mol/kg, g_1 = 0.01 mol/kg, τ = 40 min, V_3 = 0.022 mol/kg*min, $V_{g,3}$ = 0.004 mol/kg*min, K = 1.7*10^{-6} min^{-3}, φ = 0.18.

The calculated diagrams of POOH accumulation in oxidizing on air samples IPP-3 (Fig. 11, curves 1 - 3, accordingly for 120, 130 and 140 °C) are received at values **c** = g_1 = 0.01 mol/kg. Other characteristics for them are accordingly equal: 1) τ = 1 h, $V_{g,3}$ = 0.056 mol/kg*h, K = 0.003 h^{-3}; 2) τ = 0.4 h, $V_{g,3}$ = 0.115 mol/kg*h, K = 0.026 h^{-3}; 3) τ = 0.0 h, $V_{g,3}$ = 0.190 mol/kg*h, K = 0.110 h^{-3}, and the generalized formulas given in a minute time scale, look like:

τ (min) = 6.66*10^{-15} exp (E_τ/RT), E_τ = 120 kJ/mol;

$V_{g,3}$ (mol/kg*min) = 7.72*10^{8} exp (-E_g/RT), E_g = 90 kJ/mol;

K (min^{-3}) = 1.94*10^{23} exp (-E_k/RT), E_k = 234 kJ/mol.

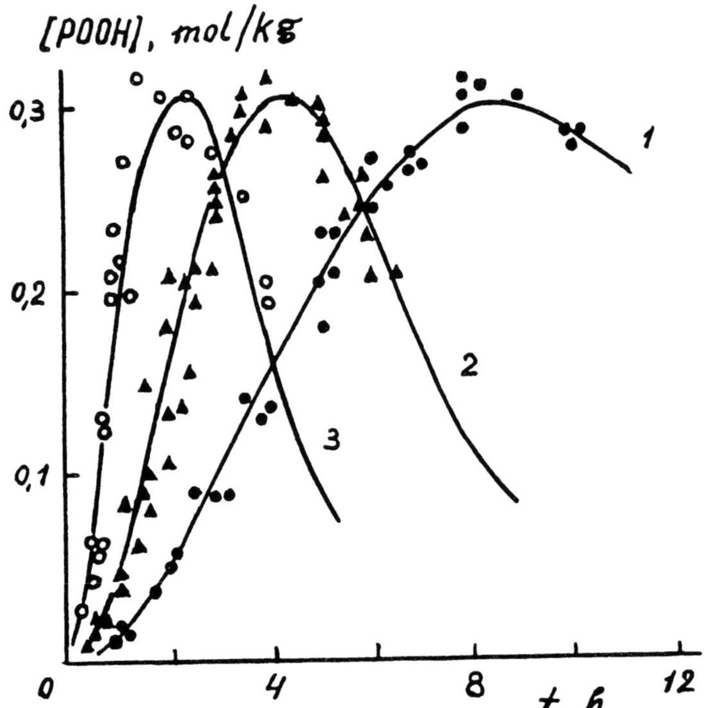

Figure 11. Calculated curves of hydroperoxide's accumulation during oxidation of IPP-3 on air at temperature 120 (1), 130 (2) and 140 °C (3).

The carried out calculations testify that the theoretical equations (11) and (15) of heterophaseous process correctly describe mutual correlation of curves of oxygen's absorption and POOH accumulation. This reveals absence of synchronism on the large interval of time and the presence of consecutive rotating of the induction periods, autoacceleration and zero order with constant rate of oxygen's absorption. During these periods the POOH yield on 1 mol of the absorbed oxygen does not change its value φ, though the POOH concentration passes through a maximum. A constancy of φ during accelerated POOH decomposition specifies that the decomposition of POOH is caused mainly by its reaction with accumulating products of reaction and that the contribution of chain reaction in this case is negligibly small.

Regarding to the catalytic reaction of POOH decomposition it should be mentioned that it has high effective energy of activation E_k. According to the scheme, the rate of catalytic reaction is determined by the interaction acts of peroxiles with end keto-groupings and acts of POOH condensation with acids (activation energy of these acts - E_5 and E_6). It is possible that these acts require rather powerful thermomechanical fluctuations, capable to open v-nanopores of micellar sponge up to the required size. Dynamics of such acts in LDPE melt correlates with higher value of E_k = 300 kJ/mol in comparison with 230 and 234 kJ/mol at APP and IPP. That can be explained with more ordered stacking of chain segments of LDPE [32, 53] (or with significant narrowness of stationary v-nanopores and higher relative gain of their volume in the act of fluctuatioinal pulsations).

The calculated curve of changes of POOH concentration on Fig. 8 - 11 in a number of cases notably deviate downwards from experimental points at higher stages of oxidation. It can be explained by the loss of acids - catalysts of POOH decomposition, not taken into account in the scheme. The process of evaporation of low-molecular weight fragments from oxidizing polymer is limited by physical diffusion at all thickness of samples, accepted to experiment, in opposite to process of oxygen's transfer [54]. Therefore, an average concentration of volatile acids depends on thickness of polymeric films and it can (in significant limits) change value of $[POOH_{max}]$.

Let's note also an observable deviation of experimental curves 3, 4 from calculated curves 5, 6 (at that stage, when the concentration of POOH is essentially reduced, passing a point of a maximum) on Fig. 2 at the large depth of oxidation. One of the reasons of this divergence can be that registered absorption of oxygen by a manometric method is deformed by accumulation of gaseous products of oxidation in an oxygen atmosphere (namely CH_4, CO, H_2) not absorbed by sorbents of water and CO_2 and not solidified at temperature of liquid nitrogen. Specified gaseous components are generated not only from oxidizing polymer, but can be formed also in result of gaseous phase oxidation of volatile products of the reaction.

8. PHENOMENOLOGICAL CONSEQUENCES

The consecutive change of macroscopic stages of reaction is a general law of oxidation of high molecular weight hyrdocarbons. Another convincing evidence is given on Fig. 12 of work [27] (the given diagrams correspond to those, as in the book [4], but differ by a variety of experimental points). Among submitted on Fig. 12 polymers are poly (ethylene), poly (propylene), copolymer of ethylene and propylene (13.5 %), poly (4-methylpenten-1), poly (styrene), rubber on the basis of synthetic poly (isoprene), poly (heptene), poly (nonene) and epoxyrubber.

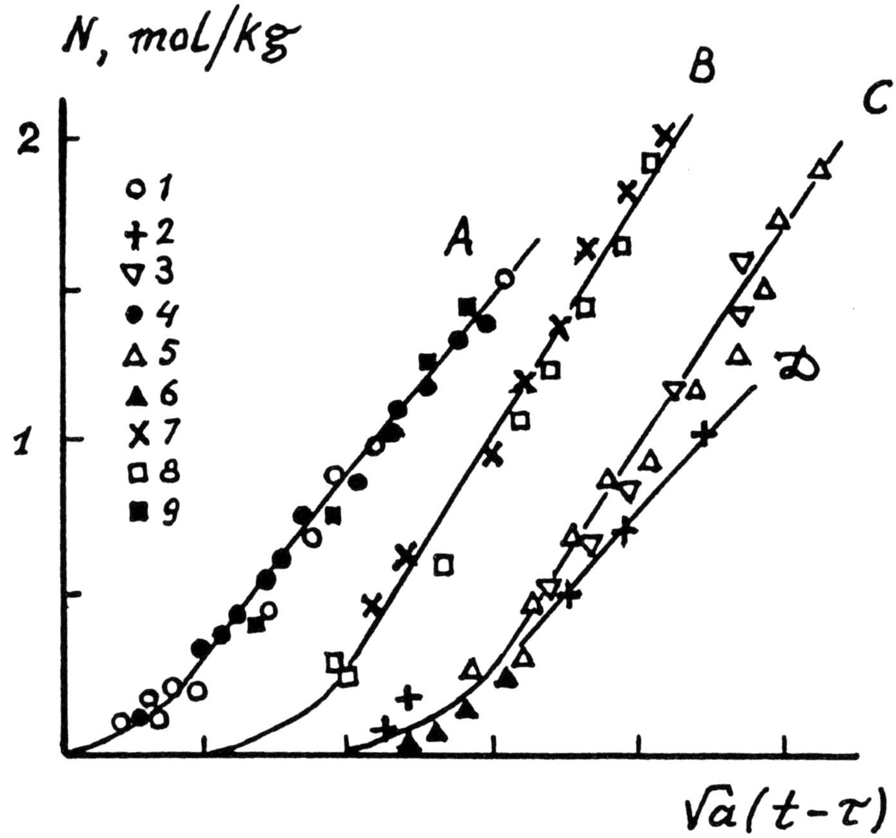

Figure 12. Dependence of quantity of the absorbed oxygen of value $a^{0.5}$ (t - τ) for oxidizing polymers: 1 – poly(ethylene) (90 °C), 2 – poly(propylene) (130 °C), 3 – copolymer of ethylene with 13.5 % propylene (130 °C), 4 – poly(4-methylpenten-1) (130 °C), 5 - rubber based of synthetic rubber (110 °C), 6 – poly(styrene) (160 °C), 7 – poly(hepten-1) (100 °C), 8 – poly(nonene-1) (100 °C), 9 - epoxyrubber (200 °C) [4, 27]. The curves B, C, D are shifted along curve A on an abscissa axis accordingly on one (B) and two (C, D) scale units, one scale unit is equal 0.5. The value of factors $a^{0.5}$, induction periods and oxygen pressure are specified in work [4].

In works [4, 27] all set of experimental points for the mentioned above polymers is put on the diagram without division (as a general field of points) and is attributed to a unique curve distinguished by a rather extended stage of autoacceleration and masking a real appearance of separate curves of each concrete polymer. Meantime, having distinguished series of experimental points for each concrete polymer (just as it's done on Fig. 12) it is easy to be convinced that the autoaccelerated stage, coming after an induction period, actually takes considerably smaller interval of time and is quickly replaced by a long stage of the zero order on hydroperoxide, reproducing the picture, considered earlier by us.

The formulas received by us for the induction periods of polyolefins' oxidation τ (min): $5.25*10^{-19}$ exp (150,000/RT) (APP), $6.66*10^{-15}$ exp (120,000/RT) (IPP-3), $4.34*10^{-15}$ exp (128,000/RT) (LDPE), can be added by the formulas, which we have calculated from the diagrams of works [27] and [55]: $3.93*10^{-12}$ exp (107,000/RT) (LDPE) and $2.5*10^{-9}$ exp (85,700/RT) (LDPE), correspondingly. This set bears evidence, on the one hand, to a wide variation of activation energy E_τ even for same chemical reagent - LDPE, on the other hand, about the occurrence of simultaneous change of preexponential factor (τ_0) according to so-called compensatory effect (CEF) [6].

It is clearly seen that all mentioned above polymers are united by common expression of CEF, which looks like (in reciprocal values of the induction period)

$$\ln(1/\tau_0) = -12.0 + 3.6*10^{-4} E_\tau$$

and does not depend on presence of the substituents on macromolecules.

The phenomenon CEF has a place for stationary rates of hydroperoxides' accumulation $V_{g,3}$ (mol/kg*min) as well: $2.1*10^6$ exp (-69,500/RT) (APP), $1.93*10^9$ exp (-90,000/RT) (IPP-3, the value is recalculated on a non-crystal phase of polymer), $1.48*10^{13}$ exp (-124,700/RT) (LDPE). The given series of the formulas can be added by calculations from the diagrams of work [55], describing oxidation of LDPE (in calculations we used established value of φ = 0.17 for LDPE): $V_{g,3}$ (mol/kg*min) equals $1.87*10^6$ exp (-76,000/RT) in area $T < T_m$ of crystallites and $4.25*10^{15}$ exp (-145,000/RT) in area $T > T_m$). In this case interrelation of values E_g and preexponential factors ($V_{g,3}^0$) is described by the equation

$$\ln(V_{g,3}^0) = -7.0 + 3*10^{-4} E_g$$

that does not depend either of presence of the substituents on macromolecules, or of change of E_g at LDPE's transition from solid state to melt.

A specific inversion of an APP site on the diagrams of straight lines of CEF can be noted: this polymer located on the top of a linear part of dependence $\ln(1/\tau_0)$ from E_τ, but moves to the bottom on a linear part of dependence $\ln(V_{g,3}^0)$ from E_g. IPP moves by a similar manner, but in smaller scale.

The polyolefins' properties expressed in the form of CEF and its transformation at transition of oxidation from first macrostage to third, do not correlate to the mechanism of homogeneous reaction of hydrocarbons. In the present moment it is obvious that the theory of heterophaseous reaction is in the beginning of its way (at a stage of construction and probation of structural - kinetic reactionary models) and does not provide a theoretical explanation of CEF yet. However that fact that the given effect has revealed as one of consequences of the heterophaseous mechanism of chain oxidation, allows to think of its interrelation with certain general properties of a micellar – spongy matrix. We believe that for the theory CEF it is necessary to study properties and ways of functioning of molecular sponge in a sufficient degree. First of all it concerns to laws of self-synchronization and cooperation of fluctuations of chain parts forming a fringe layer around nanopore [28, 29]. The thermal fluctuations of parts of a fringe layer converge on of nanopore's wall and nanopore increases its volume at concurrence on it of oscillatory energy from several chains. As a result of it reactionary or migration centers adhered to molecular sponge continuously experiencing frequent modulation of its energetic and steric state. It is possible that CEF expresses dependence of nanopore's dynamics (localizing into itself the key chemical acts) of a thermomechanical condition of spongy micelles. In particular, the number, length or arrangement of side branches on the macromolecule's backbone, and also degree of flexibility and tension of polymeric chains can anyhow affect on mobility of sponge's parts, frequency and amplitude of nanopore's pulsations. In that case the CEF's behavior at transition from the induction period to stationary oxidation can reflect a qualitative change of the fluctuation acts serving formation of free radicals in s-zones.

The facts, combined with CEF, specify an inconsistency of the widespread opinion that the rate of reaction should necessarily grow at increase of molecular-segmental mobility. Really, rate $V_{g,3}$ at 130 °C equals 2.0*10^{-3} and 1.0*10^{-3} mol/kg*min, accordingly in melt of APP and LDPE, but 4.0*10^{-3} in solid IPP-3.

Is remarkable that IPP oxidizing faster at 130 and 150 °C differs from LDPE in 6 times by smaller rate of macromolecule's cleavage, determined by a viscometry [41]. Our heterophaseous scheme explains this not explained earlier [41] fact with presence at IPP of a competitive way of disintegration of polymeric alkoxyle, formed in a epoxy reactionary branch, with formation of free methyl radical. LDPE can not follow this mechanism and its medial alkoxyles proceed mainly with macromolecule's cleavage.

Let's remind that the statistical acts of macromolecule cleave (without loss of end fragments) proceed in s-zones of spongy micelles as well. Their contribution to total process of oxidizing destruction is insignificant by virtue of a small mass fraction of polymeric s-segments. Meantime, in the special experiments they can determine physical essence of the process, for example, providing the kinetic first order on the initiator in process of chemically initiated relaxation of a tense applied to crosslinked polymers [56]. The first order of process on the initiator means that the mechanical durability of polymeric samples in experimental conditions is determined by degree of damage of continuous spatial paracrystalline skeleton first, *i.e.* by degree of conversion of that process, which proceeds directly in s-zones of spongy micelle [21].

One more important remark concerns to factor **c** = 0.025 mol/kg, found for films IPP-1 and IPP-2. Its value precisely equals to a constant **c**, established earlier in experiments on anaerobic decomposition of hydroperoxide in preliminary oxidized films IPP-1 [34]. It happened because POOH decomposition in this polymer in absence of oxygen is caused by chain heterophaseous reaction, whose rate depends not only on POOH concentration in v-zones of spongy micelles, but also of POOH concentration in s-zones, and is proportional to value $g_s^{0.5}$. The initial concentration g_s in anaerobic reaction is determined by the equation (7) of oxidizing processes, therefore constant **c** is characteristic of both oxidizing and anaerobic processes at the same time. Thus, the size **c** = 0.025 mol/kg being a parameter of an autoaccelerated stage of the IPP-1 oxidation, fixes by itself a real algebraic interrelation between rates of two heterophaseous processes proceeding in essentially differing conditions: in absence of oxygen and at its active participation in transformations.

Let's also emphasize that the third macroscopic stage for process of polymeric hydrocarbon's oxidation is not last, because sooner or later the content of CH-groups and POOH in s-zones should be lowered in very significant degree. All this in combination to simultaneously proceeding change of molecular-sponge micelles' structure inevitably makes the equation (7) invalid and will change reaction's kinetics. At the same time, the material, considered by us, allows to conclude that the heterogeneous-heterophaseous reactionary model considerably expands limits of the kinetic description of oxidizing transformations of polymeric hydrocarbons. The additional important evidence in favor of such conclusion gives a material of the following chapter of present article.

9. The Peculiarities of Oxidation of Poly (Styrene) Melt

Kinetic curves of oxygen's absorption and POOH accumulation in oxidizing melt of poly (styrene) (PS) given in work [57] (T = 180 – 210 °C, p = 26.7 kPa), qualitatively repeat a situation noticed with polyolefins (Fig. 13, 14). It is quite natural for carbochain polymer with reactive CH-bonds, which is PS, having same molecular-spongy organization of polymeric chains, as well as all non-crystalline polymers [38]. At the same time, in a quantitative sense PS differs in a hundred times by lower concentration of formed POOH (Fig. 14).

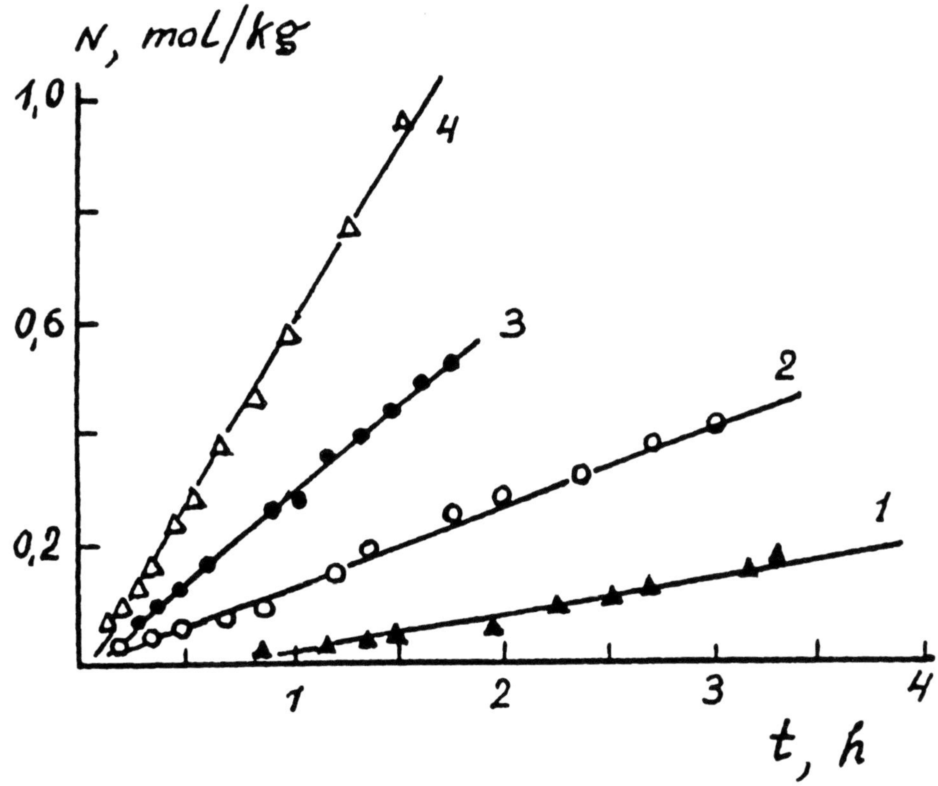

Figure 13. Curves of oxygen's absorption during oxidation of PS melt at temperature 180 (1), 190 (2), 200 (3) and 210 °C (4), p = 26.7 kPa [57].

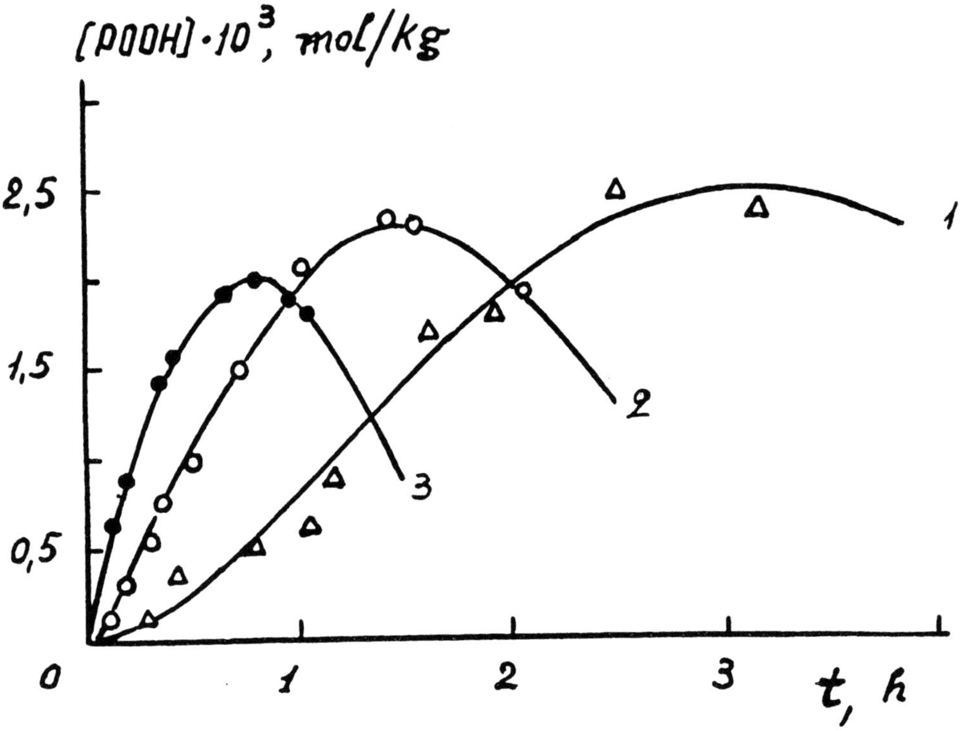

Figure 14. Calculated curves of hedroperoxide's accumulation during oxidation of PS melt at temperature 180 (1), 190 (2), and 200 °C (3) [57].

The calculations, carried out by us, have shown, that the experimental data of work [57] is well coordinated to the theoretical equations (11), (15) of heterophaseous process (the curves on Fig. 13, 14 are calculated). In this case factor **c** = 0.0002 mol/kg and concentration g_1 = 0.00015 mol/kg are constant in an interval of temperature 180 –210 °C and other characteristics have the following meanings: 1) τ = 0.4 h, V_3 = 0.06 mol/kg*h, $V_{g,3} = 1.35*10^{-3}$ mol/kg*h, K = 0.07 h^{-3}, $\varphi = 2.25*10^{-2}$ (180 °C, Fig. 13, 14, curves 1); 2) τ = 0.05 h, V_3 = 0.14 mol/kg*h, $V_{g,3} = 2.35*10^{-3}$ mol/kg*h, K = 0.44 h^{-3}, $\varphi = 1.68*10^{-2}$ (190 °C, Fig. 13, 14, curves 2); 3) $\tau \approx$ 0.00 h, V_3 = 0.30 mol/kg*h, $V_{g,3} = 3.8*10^{-3}$ mol/kg*h, K = 2.8 h^{-3}, $\varphi = 1.27*10^{-2}$ (200 °C, Fig. 13, 14, curves 3); 4) $\tau \approx$ 0.00 h, V_3 = 0.61 mol/kg*h (210 °C, Fig. 14, curve 4).

The appropriate generalized formulas calculated for a minute time scale, look like:

$$V_3\ (\text{mol/kg*h}) = 9.7*10^{12} \exp(-E/RT),\ E = 138.5\ \text{kJ/mol};$$

$$V_{g,3}\ (\text{mol/kg*h}) = 3.6*10^{5} \exp(-E_g/RT),\ E_g = 88.3\ \text{kJ/mol}$$

$$K\ (\text{min}^{-3}) = 7.6*10^{10} \exp(-E_k/RT),\ E_k = 151.0\ \text{kJ/mol}.$$

Comparing these results with the appropriate characteristics of polyolefins we note 100 times lower values of factor **c** and concentration g_1. Oxidizability of PS in equal with polyolefins conditions is much lower as well. For example, poly (styrene) values V_3 and $V_{g,3}$ recalculated on 130 °C are $1.08*10^{-5}$ and $1.28*10^{-6}$ mol/kg*min respectively. The yield of hydroperoxide, which practically does not depend of temperature for LDPE (0.167), is reduced for PS at its increase, φ: $11.8*10^{-2}$ (130 °C); $2.25*10^{-2}$ (180 °C); $1.68*10^{-2}$ (190 °C); $1.27*10^{-2}$ (200 °C). It is caused by unequal values of effective activation energy of processes of oxygen's absorption E = 138.5 kJ/mol and the peroxide's accumulation E_g = 88.3 kJ/mol (they are equal for LDPE).

Low values of factor **c** and the concentration g_1 for hydroperoxide on exit from the induction period reflect high fluctuational dynamics of molecular sponge in PS melt. The expression of factor **c** = $k_{0s}k_{-f}/k^f k_f$ testifies to it, which denominator corresponds to multiplication of fluctuational rate constants (it in 100 times bigger in value than at LDPE melt). Thus, it can be supposed that the much higher frequency of pulsations of nanopores in PS sponge provides higher speed of migration of OOH-groups from v-zones of spongy micelles in s-zones, as well as high rate of radicalless bimolecular reaction ($POOH_f$ + $POOH_s$) and, accordingly, realization of the equations (7), (8), (11), (13).

Proceeding from high flickering activity of nanopores in PS molecular sponge, simultaneous increase of rate of square-law cleavage of reactionary v-chains may be expected. In this connection we shall note that written earlier stage for polyolefins

$$2POO^\bullet \xrightarrow{k_t} \text{termination}$$

obviously does not reflect influence of the structural factor, keeping formally kind of homogeneous reaction. The given simplified approach has not caused the special consequences at the kinetic description of process of polyolefin's oxidation, however by consideration of a wider complex of objects it is required to detail the mechanism of termination of reactionary chains.

The more detailed scheme of interaction of polymeric peroxiles in molecular sponge should reflect the acts of fluctuational excitation and deactivation of nanopores, owning by peroxiles, acts of appearance of peroxiles in the same nanopore and competition of the acts of interaction and yield of peroxiles from same nanopore:

$$POO^\bullet \underset{k_{-t,1}}{\overset{k_{t,1}}{\leftrightarrow}} POO_f^\bullet, \qquad POO_f^\bullet + POO^\bullet \xrightarrow{k_{t,2}} (POO^\bullet + POO^\bullet)_f$$

$$(POO^\bullet + POO^\bullet) \xrightarrow{k_{t,3}} termination, \qquad (POO^\bullet + POO^\bullet)_f \xrightarrow{k_{t,4}} 2POO^\bullet.$$

Applying a condition of stationary concentration to fluctuationally exited peroxiles and radical pairs it is easy to express a rate of termination of reactionary chains:

$$V_t = -d[POO^\bullet]/dt = 2k_{t,1}k_{t,2}k_{t,3}[POO^\bullet]^2/(k_{-t,1} + k_{t,2}[POO^\bullet])(k_{t,3} + k_{t,4})$$

The deduced expression transforms in limiting cases either to

$$V_t'' = [2k_{t,1}k_{t,2}k_{t,3} / k_{-t,1}(k_{t,3} + k_{t,4})][POO^\bullet]^2 = k_t''[POO^\bullet]^2 \quad (17)$$

(at high probability of deactivation of fluctuationally exited peroxiles $k_{-t,1} > k_{t,2}[POO^\bullet]$), or to

$$V_t' = [2k_{t,1}k_{t,3} /(k_{t,3} + k_{t,4})][POO^\bullet] = k_t'[POO^\bullet] \quad (18)$$

adequate to higher frequency of occurrence of two peroxiles in one nanopore ($k_{-t,1} < k_{t,2}[POO^\bullet]$).

The equation (17) expresses square-law dependence of termination rate of peroxiles' concentration however (as to its essence) describes heterogeneous process. In its turn, the equation (18) describes heterogeneous linear termination of reactionary chains, impossible at occurrence of reactionary equivalent peroxiles in homogeneous media.

It is possible to conclude about presence in the given process of square-law termination of reactionary chains according to the equation (17), taking into account, that oxidation kinetics of PS is described by the equations (11), (15) for heterogeneous reaction. The latter allows to connect low rate of PS oxidation (compare to polyolefins) with higher value of a termination rate constant $k_t'' = 2k_{t,1}k_{t,2}k_{t,3} / k_{-t,1}(k_{t,3} + k_{t,4})$. The higher value of k''_t can be caused by a high fluctuation constant $k_{t,1}$ and low constant $k_{-t,1}$ of collapsing increased in volume v-nanopores up to the initial size.

Fluctuational pulsation of nanopores occur at simultaneous compression or expansion of zones of macromolecular sponge, surrounding a nanopore, that anyhow depends of an arrangement and spatial movements of phenyl rings adhered to the macromolecules' backbone. Phenyl rings, occupying a place in nanopores' volumes, increase probability of expansion and reduce probability of nanopores' collapse, thus cooperating fluctuational oscillations of sponge's surrounding chains. For the same reason constant $k_{t,2}$ of OOH-groups' capture by flickering nanopores can also increase.

At the same time, phenyl rings do not change mechanisms of heterophaseous initiation and termination of reactionary chains in zones of a PS matrix and consequently expressions (7), (13) for stationary concentration of peroxiles remain unchanged. Qualitative kinetic conformity of oxidation processes of PS and polyolefins allows to write reactions of formation of PS oxidation products using a method of analogy. In particular, end keto-groups can be formed by the following way:

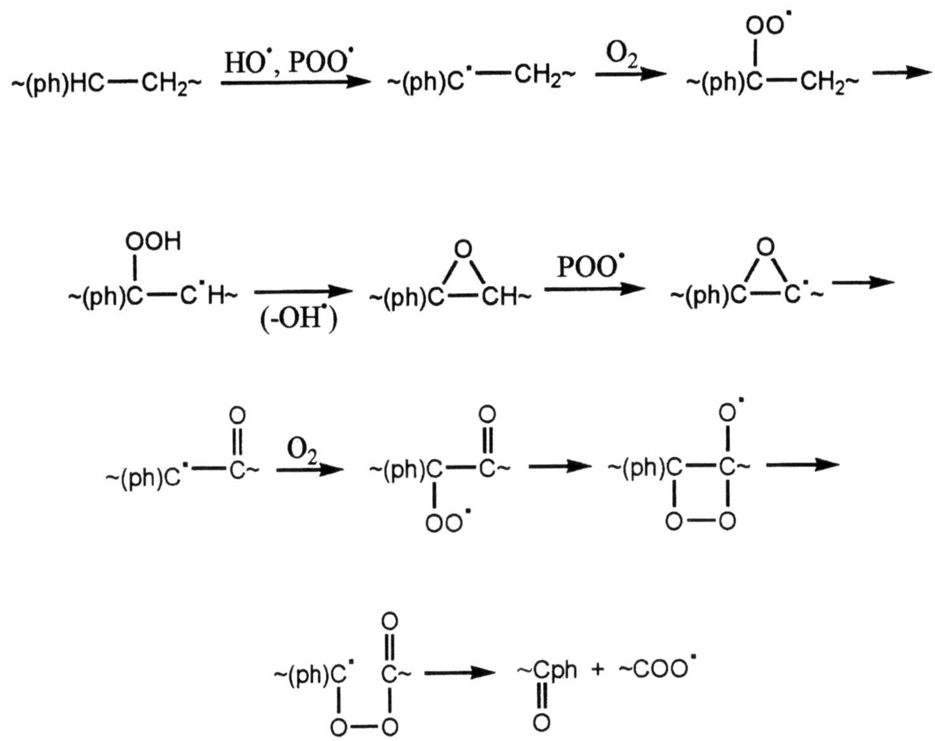

$$\sim COO^{\bullet} + PH \longrightarrow \sim COOH + P^{\bullet}, \quad \sim COO^{\bullet} \longrightarrow CO_2 + \sim CH_2CH^{\bullet}ph$$

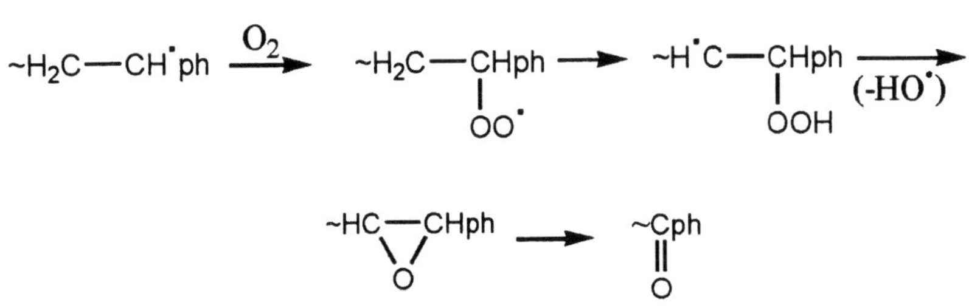

They appears to be a source low molecular weight benzaldehyde and benzoic acid:

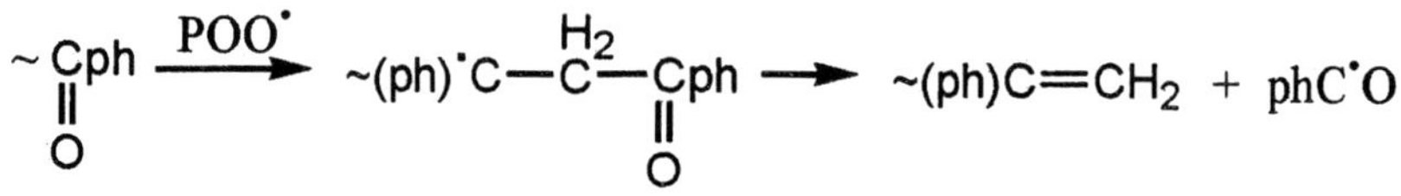

$$phC^{\bullet}O + PH \longrightarrow phCHO + P^{\bullet}$$

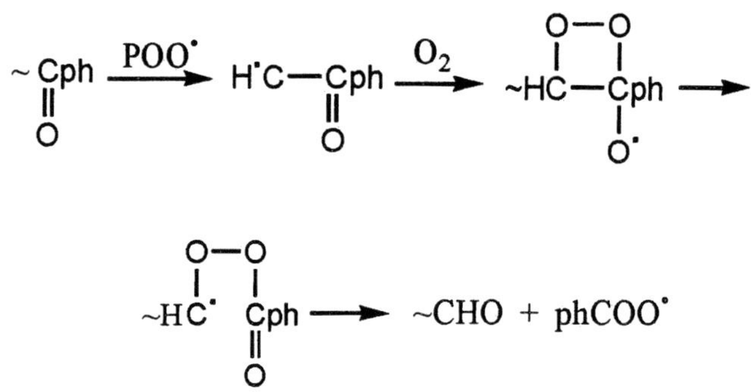

$$phCOO^{\bullet} + PH \longrightarrow phCOOH + P^{\bullet}, \quad P^{\bullet} + O_2 \longrightarrow POO^{\bullet}$$

Benzoic acid, accumulating in PS melt, catalyses POOH decomposition according to the equation (15), and, having higher acidic properties in comparison with alkylcarbonic acids, it is the more active catalyst because of higher condensation rates with POOH and disassociation of intermediate alkylaromatic perester.

Formation of acetophenone can be explained by oxidation of formed styrene:

$$\sim CH_2C^{\bullet}Hph \longrightarrow \sim CH_2CH^{\bullet}ph + CH_2{=}CHph$$

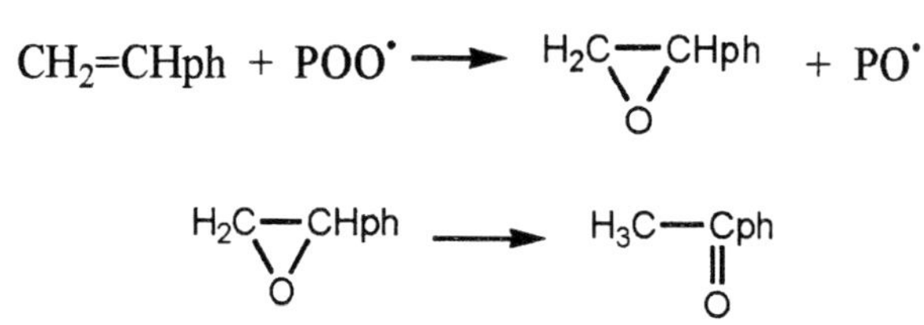

$$PO^{\bullet} + PH \longrightarrow POH + P^{\bullet}$$

or by reactions of epoxy compounds with simultaneous formation of unsaturated conjugated groups:

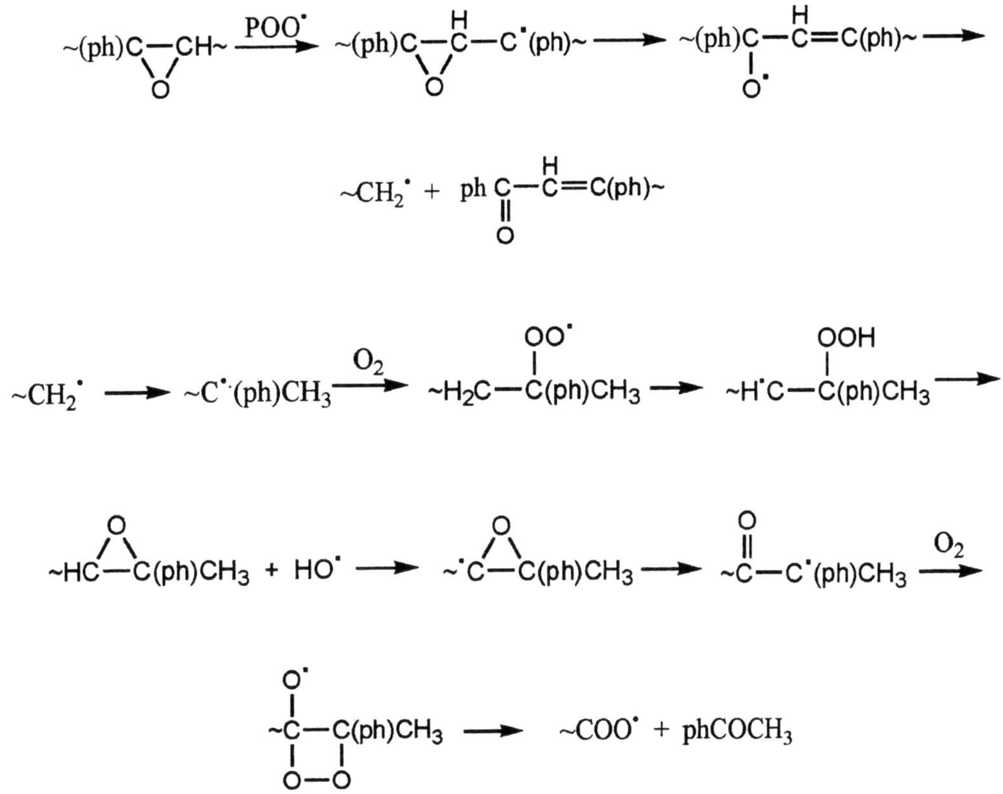

Based on its analogy with polyolefins the scheme of PS chemical transformations can be expanded not changing expressions (7) and (13). At the same time, the method of analogy does not allow to explain an important difference of PS of LDPE, related to decreased yield $\varphi = V_{g,3}/V_3$ of hydroperoxide at increased oxidation temperature.

Reaction rates of POOH and epoxy rings formation in oxidizing LDPE are determined by removal rates of hydrogen atoms by peroxiles, accordingly, with constants k_2 and k_3 at practically identical activation energy E_2 and E_3. Owing to this the equality $E_g = E$ takes place and factor of an hydroperoxide yield $\varphi = k_2[POO^\bullet]/(k_2 + \eta k_3)[POO^\bullet]$ practically does not depend on temperature.

Oxidizing PS differs by an inequality $E > E_g$, which is responsible for decrease of yield φ of hydroperoxide at increased temperature. The given fact bears evidence about the existence of one more way of transformations of peroxiles, competing with POOH and epoxy rings formation. This way should increase oxygen consumption rate at increased temperature simultaneously with POOH and epoxides formation not changing the mechanism of reactionary chains termination, but differing by higher activation energy for E_2 (E_3). Its presence should be taken into account in terms of hydroperoxide's yield as an additional component in denominator's structure:

$$\varphi = k_2[POO^\bullet]/(k_2 + \eta k_3 + k_{ph})[POO^\bullet]$$

where k_{ph} – rate constant of a new process. Most probable in this respect is the reaction with attack of oxygen peroxide on adjacent phenyl ring with subsequent leave of free phenoxyl:

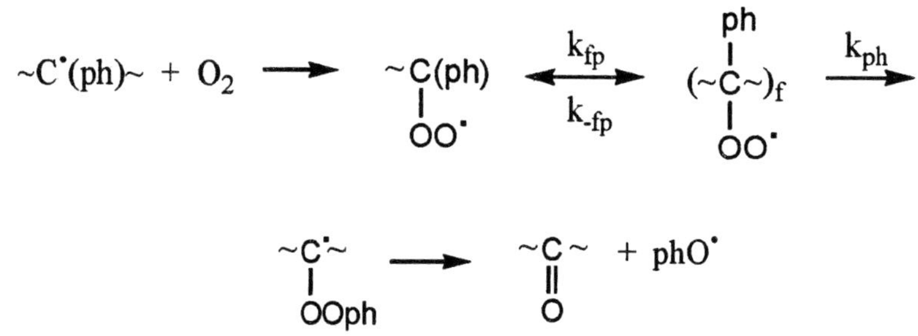

The free space, necessary for the given process, can be provided during of nanopores' volumetric pulsations.

Formed by such way free phenoxyl continues a reactionary oxidation chain, turning into a phenol owing to high enough reactivity and due to high temperature of PS oxidation. The phenol formation at oxidation of PS was experimentally established in work [58].

10. Conclusion

The purpose of present article is to continue structural - kinetical modeling of chain reactions proceeding in a non-crystalline polymeric matrix, on an example of polymeric hydrocarbons. A model of molecular-sponge micelle - major structural unit of non-crystallinic polymer, including zones with non-uniform on the size and on physico-dynamic properties nanopores became a morphological basis for this purpose. Keeping qualitatively identical form in solid (glass), devitrificated and melt conditions, micelle is a source of many important properties of polymers [28, 29, 38, 39, 59].

Developed reactionary kinetical model, taking matrix heterophasiety of nanostructural scale in account, already has earlier allowed to explain many features of chain free-radical transformations, which are not described within the framework of homogeneous reactionary model [15, 16, 18 - 22, 38, 60].

The high efficiency of heterophaseous reactionary model is shown in present article in the regard of non-crystal polymeric hydrocarbons with the assistance of the kinetic analysis of alternating in time macroscopic stages of process of oxidation.

The results of the analysis change existing representations about the mechanism and kinetics of polymer's autooxidation and, opening a fundamental role of micellar nanophase in oxidizing process, bear new questions on concrete details of functioning molecular-chain sponge in chemical transformations.

References

[1] Pudov, V.S.; Gromov, B.A.; Sklyarova, E.G. *Neftekhimiya* **1963**, *3(5)*, 743.

[2] Pudov, V.S.; Neiman M.B. In book "*Aging and stabilization of polymers*" (Ed. Kuzminskii, A.S.); Chemistry: Moscow, 1966; p. 5.
[3] Reich, L., Stivala, S.S. *Autoxidation of Hydrocarbons and Polyolefins. Kinetics and Mechanisms*; Marcel Dekker: New York, 1969.
[4] Slyapnikov, Yu.A.; Kiryushkin, S.G.; Mar'in, A.P. *Antioxidizing stabilization of polymers*; Chemistry: Moscow, 1986.
[5] Popov, A.A.; Rapoport, N.Ya.; Zaikov, G.E. *Oxidation of the oriented and stressed polymers*; Chemistry: Moscow, 1987.
[6] Emanuel, N.M.; Buchachenko, A.L. *Chemical physics of molecular destruction and stabilization of polymers*; Science: Moscow, 1988.
[7] Roginskii, V.A. *Phenolic Antioxidants. Reactionary ability and efficiency*; Science; Moscow, 1988.
[8] Denisov, E.T. *Oxidation and destruction of carbochain polymers;* Chemistry: Leningrad, 1990.
[9] Vago, J.; Fodor, Z.; Nagy, T.T.; Iring, M.; Kelen, T.; Tüdõs, F. *Polym. Bull.* **1983,** *10(7/8)*, 559.
[10] Kelen, T.; Iring, M.; Tüdõs, F. *Eur. Polym. J.* **1976,** *12(1)*, 35.
[11] Iring, M.; Kelen, T.; Tüdõs, F. *Makromol. Chem.* **1974,** *175(2)*, 467.
[12] Denisov, E.T. *Uspekhi Khimii* **1978,** *47(6)*, 1090.
[13] Emanuel, N.M.; Buchachenko, A.L. *Chemical physics of aging and stabilization of polymers;* Science: Moscow, 1982.
[14] Bartenev, G.M.; Zelenev, Yu.V. *Physics of polymers*; Chemistry: Leningrad, 1976.
[15] Mikheev, Yu.A.; Guseva, L.N.; Zaikov.G.E. *Vysokomol. Soedin.* **1997,** *39(6)*, 1082.
[16] Mikheev, Yu.A.; Guseva, L.N.; Zaikov.G.E. *Uspekhi Khimii* **1997,** *66(1)*, 3.
[17] Slyapnikov, Yu.A. *Khim. Fizika* **1995,** *14(10)*, 48.
[18] Mikheev, Yu.A.; Guseva, L.N. *Khim Fizika* **1987,** *6(9)*, 1259.
[19] Mikheev, Yu.A.; Guseva, L.N.; Toptygin, D.Ya. *Vysokomol. Soedin. Ser. A* **1989,** 31.
[20] Mikheev, Yu.A.; Guseva, L.N. *Khim Fizika* **1992,** *11(7)*, 964.
[21] Mikheev, Yu.A.; Guseva, L.N. *Khim Fizika* **1993,** *12(8)*, 1081.
[22] Mikheev, Yu.A. *Khim Fizika* 1999, 18(8), (in press).
[23] Bogaevskaya, T.A.; Gromov, B.A.; Miller, V.B.; Monakhova, T.V.; Shlyapnikov, Yu.A. *Vysokomol. Soedin, Ser. A* **1972,** *14(7)*, 1552.
[24] Monakhova, T.V.; Bogaevskaya, T.A.; Gromov, B.A.; Shlyapnikov, Yu.A. *Vysokomol. Soedin. Ser. A* **1974,** *16(2)*, 91.
[25] Bogaevskaya, T.A.; Monakhova, T.V.; Shlyapnikov, Yu.A. *Vysokomol. Soedin. Ser. A* **1978,** *20(6)*, 465.
[26] Iring, M.; Laszlo-Hedvig, Zs.; Tüdõs, F.; Kelen, T. *Polym. Degrad. Stab.* **1983,** *5(6)*, 467.
[27] Emanuel, N.M. *Vysokomol. Soedin. Ser. A* **1985,** *27(7)*, 1347.
[28] Mikheev, Yu.A.; Guseva, L.N.; Zaikov.G.E. *Vysokomol. Soedin. Ser. A* **1999,** *41(2)*, 346.
[29] Mikheev, Yu.A.; Zaikov.G.E. *Vysokomol. Soedin. Ser. A* **1999,** *41(5)*, 852.
[30] Pudov, V.S. Candidate of Chemical Sciences, Dissertation, USSR, Moscow, 1963.
[31] Monakhova, T.V. Candidate of Chemical Sciences, Dissertation, USSR, Moscow, 1976.
[32] Aandrianova, G.P. *Physico-Chemistry of Polyolefins*; Chemistry: Moscow, 1974.

[33] Mikheev, Yu.A.; Guseva, L.N.; Mikheeva, L.E.; Kalantsova, S.V. *Khim. Fizika* **1993**, *12(4)*, 528.
[34] Mikheev, Yu.A.; Guseva, L.N. *Khim. Fizika* **1996**, *15(9)*, 131.
[35] Shibryaeva, L.S.,; Kiryushkin, S.G.; Zaikov, G.E. *Int. J. Polym. Mater.* **1992**, *16(1-4)*, 71.
[36] Aleskovskii, V.B.; Korsakov, V.G. *Physico-chemical basis of a rational choice of active materials*; Publishing house of the Leningrad university: Leningrad, 1980.
[37] Iring, M.; Kelen, T.; Fodor, Zs.; Tüdõs, F. *Polym. Bull.* **1982**, *7(9/10)*, 489.
[38] Mikheev, Yu.A.; Zaikov, G.E. *Khim. Fizika* **1997**, *16(11)*, 62.
[39] Mikheev, Yu.A.; Guseva, L.N.; Zaikov, G.E. *J. Appl. Polym. Sci.* **1998**, *67*, 1693.
[40] Loftus, J.; Satterfield, C.N. *J. Phys. Chem.* **1965**, *69(3)*, 909.
[41] Barabas, K.; Iring, M.; Laszlo-Hedvig, Zs.; Kelen, T.; Tüdõs, F. *Eur. Polym. J.* **1978**, *14(6)*, 405.
[42] Iring, M.; Laszlo-Hedvig, Zs.; Barabas, K.; Kelen, T.; Tüdõs, F. *Eur. Polym. J.* **1978**, *14(6)*, 439.
[43] Barabas, K.; Iring, M.; Kelen, T.; Tüdõs, F. *J. Polym. Sci., Symp.* **1976**, *57*, 65.
[44] Yuriev, V.M.; Pravednikov, A.N.; Medvedev, S.S. *J. Polym. Sci.* **1961**, *55(11)*, 353.
[45] Antonovskii, V.L.; Denisov, E.T.; Solntseva, L.V. *Kinetika I Kataliz* **1966**, *7(3)*, 409.
[46] Beachell, H.C.; Beck, D.L. *J. Polym. Sci. A.* **1965**, *3(2)*, 457.
[47] Faulkner, D.L. *Polym. Eng. Sci.* **1982**, *22(8)*, 466.
[48] Ryshavy, D.; Balaban, L.; Slavik, V.; Ruzha, Ya. *Vysokomol. Soedin.* **1961**, *3(7)*, 1110.
[49] Cheng, H.N.; Schilling, F.C.; Bovey, F.A. *Macromolecules* **1976**, *9(2)*, 363.
[50] Manyasek, Z.; Berek, D.; Michko, M.; Lazar, M.; Pavlinets, Yu. *Vysokomol. Soedin.* **1961**, *3(7)*, 1105.
[51] Latyshkaeva, I.G.; Bogaevskaya, T.A.; Belov, G.P.; Shlyapnikov, Yu.A. *Vysokomol. Soedin.* **1977**, *19(7)*, 531.
[52] Dyakonov, V.P. *The reference book on calculations on microcalculators*; Science: Moscow, 1986.
[53] Antipov, E.M.; Ovchinnikov, Yu.K.; Rebrov, A.V.; Belov, G.P.; Markova, G.S.; Bekeev, R.F. *Vysokomol. Soedin.* **1978**, *20(8)*, 1727.
[54] Iring, M.; Kelen, T.; Tüdõs, F. *Eur. Polym. J.* **1975**, *11(9)*, 631.
[55] Iring, M.; Kelen, T.; Tüdõs, F. *Polym. Bull.* **1986**, *15(6)*, 531.
[56] Tobolsky, A.V.; Norling, P.M.; Frick, N.H.; Yu, H. *J. Am. Chem. Soc.* **1964**, *86*, 3925.
[57] Goldberg, V.M.; Esenin, V.N.; Krasotkina, I.A. *Vysokomol. Soedin.* **1977**, *19(8)*, 1720.
[58] Dulog, L.; David, K.-H. *Makromol. Chem.* **1971**, *145*, 67.
[59] Mikheev, Yu.A.; Pustoshnaya, L.S.; Zaikov, G.E. *Int. J. Polym. Mater.* **1996**, *33*, 141.
[60] Mikheev, Yu.A.; Guseva, L.N.; Zaikov, G.E. *Int. J. Polym. Mater.* **1997**, *35*, 193.

Chapter 8

GENERALIZATION OF POLYMERS INTERACTION WITH ORGANIC SOLVENTS PROCESSES BY MEANS OF LINEAR POLYPARAMETER EQUATIONS. 2. SWELLING OF POLYBUTADIENE AND BUTYLRUBBER

R. Makitra[1], E. Sagladko[2], A. Turovskiy[1] and G. Zaikov[2]

[1]Department of Physical Chemistry of Fuels, Pisarshevski`s Institute of Physical Chemistry Ukrainian National Academy of Sciences, Lviv Naukova Str 3a 79053

[2]N. Emanuel Institute Biochemistry and Physics, Russian Academy of Sciences, 4 Kosygin Street, Moscow, 117334, Russia

Key words: swelling, organic solvents, polybutadiene, butylrubber, molar volume, solvation

Molar volume of organic solvents is the factor attributing their penetration into the structure of butylrubber and polybutadiene. However a quantitative dependence between degree of swelling and physico-chemical characteristics of solvents may be obtained under the condition of additional taking into account of their ability to nonspecific und specific solvation by means of linear multiparameter equation.

As the present descriptions of polymer swelling process in organic solvents have only a semiquantitative character, in particular these based on the theories of regular solutions or Flory theory [1, 2] as well these taked in quality of determining factor the solvent molecule dimensions (its molar volume V_M) [3], we assume that in result of complexity of this process it is impossible to describe it by means of only one some factor.

On the degree and rate of swelling will be influenced both molecule dimensions and the following solvation process, occurred between solvent and polymer in its structure, appointed in result of specific and non-specific interaction of system components; the significance of their influence will be attributed by structure and properties of components.

For receiving of an adequate numerical connection between the swelling characteristics and solvent properties one should simultaneously take into account the all realised into system processes which is possible in the first approximation on the basis of Linear Free

Energies Principle (LFE) in assumption of the interindependence of separate solvation processes and permissibility of summation of their effects.

In the previous paper [4] we shown that the polyethylene swelling data (the equilibric amounts absorbed solvent in mole procent S) may be quantitatively connected with physico-chemical characteristics of organic solvents by means of the six-parameter equation (1):

$$\lg S = a_0 + a_1 \frac{n^2-1}{n^2+2} + a_2 \frac{\varepsilon-1}{2\varepsilon+1} + a_3 B + a_4 E_T + a_5 \delta^2 + a_6 V_M, \qquad (1)$$

This equation takes into account both possible solvation processes going between polymer and solvent, characterized by its polarizability, polarity, basicity and electrophilicity (terms 1-4) and the expenditure of energy on the overcoming of liquids cohesion energy forcies, proportional to its solubility parameters square δ^2 as well as the influence of molar volume (the dimension of molecule) V_M. In the polyethylene case this last factor which makes difficulter the liquid diffusion into polymer is deciding and diminishing of S. But for the receiving of one adequate connection there is necessary a supplementary account of both the factors of non-specific solvation, which raises the S and the specific which it diminishes probably in the result of solvophobic properties of non-polar polyethylene.

As the swelling process substantially depends from polymer character we presumed that it will be expedient to verify the applicability of the Eq. 1 for the generalization of swelling data in organics solvents for the many other polymers of various nature accompanied by the analysis of changes of equation separate terms and their validity i.e. the solution characteristics influence on the S values. In this paper we consider the generalization of polybutadiene and butylrubber swelling data as these are near on their character to polyethylene.

In the Table 1 we give the data of the swelling of saturated poly-butadiene samples in 13 solvents based on the experimental results from [5]. As these equilibric swelling data are given in volumes of solvent imbibed after 5 days at 25°C by 1 volume of polymer, these results we recalculated in the moles of absorbed solvent. The density of investigated polybutadiene was equal 0.971 and the middle molecular weight of hydrocarbonic sections between the sewing links is 5600 i.e. these are realized across at ca. 400-th carbon atom.

The calculation for the all 13 solvents from the Table 1 gives a equation with the unsatisfactory degree of connection – the value of multiple coefficient of correlation R is equal only 0.914. However the exclusion from consideration accordingly to IUPAC recommendations the most deviating results for nitrobenzene leads to the Eq. 2 with the excellent correlation:

$$\lg S = -1.16 + (3.24 \pm 0.63)\, f(n^2) + (5.08 \pm 0.83)\, f(\varepsilon) - (1.85 \pm 0.43)\, 10^{-3} B$$
$$- (0.10 \pm 0.02)\, E_T - (4.03 \pm 0.84) 10^{-3}\, \delta^2 - (1.56 \pm 0.77) 10^{-3}\, V_M \qquad (2)$$

with R=0.996, the middle square error s±0.02 and the follow values of the pair correlation coefficients of the function S with the individual parameters r_i: 0.707; 0.702; 0.014; 0.570; 0.598 and 0.703. The adequacy of both the equation (2) and its R value is confirmed accordingly [6] by confrontation of calculated for these Fisher`s parameters values with tabulated values for corresponding number at freedom degrees at reliability α=0.95.

Table 1. Experimental (According to [3]) and Calculated by Means of Eq. 3 Amounts of Absorbed Solvents at the Swelling of Samples of Polybutadiene

N	Solvation	$S_{M(exp)}$	$lgS_{M(exp)}$	$lgS_{M(calc)}$	ΔlgS
1	Hexane	0.011	-1.9731	-1.9325	0.0405
2	Heptane	0.010	-1.9983	-1.9483	0.0499
3	Octane	0.013	-1.8831	-1.9729	-0.0898
4	Cyclohexane	0.028	-1.5494	-1.5486	0.0886
5	Tetrachloromethane	0.045	-1.3460	-1.3861	-0.0401
6	Ethylbenzene	0.031	-1.5123	-1.4829	0.0294
7	p-Xylene	0.030	-1.5280	1.4998	0.0281
8	Toluene	0.037	-1.4366	-1.4175	0.0191
9	Benzene	0.039	-1.4060	-1.4365	-0.0305
10	Trichloromethane	0.056	-1.2525	-1.2534	-0.0855
11	Nitrobenzene	0.010	-2.0186	-0.4917	-0.0153
12	Dioxane	0.015	-1.8099	-1.8167	-0.0068

Like as in polyethylene case [4] the growth of both solvent molecule dimensions V_m and its ability to specific interaction depress the solvent penetration into polybutadiene structure; on the contrary the ability to nonspecific solvation favours the penetration.

As neither the standard deviations of individual regression coefficients nor the pair correlation coefficient values with lg S_m did not attribute their influence, the validity of separate equation terms may be defined, in according with recommendations of [6], in the way of these exclusion in turn and calculation of R values for every of these equation with the lesser number of terms. Thus we stated the real insignificance of cohesion energy density δ^2 and basicity B terms for the description of swelling process by means of Eq. (2) and so the connection between the physico-chemical solvent properties and their absorption by polybutadiene samples may be satisfactorily described by follow four-parameter equation:

$$lgS=-5.12+(5.77\pm0.61)f(n^2)+(6.39\pm0.71)f(\varepsilon)-(0.11\pm0.01)E_T-(4.63\pm0.76)10^{-3}V_M \quad (3)$$

R = 0.988; s ± 0.04

However, the subsequent lowering of parameters number is inadmissible as the exclusion of V_m leads to the three-parameter equation $lgS = f(n^2, r, E_T)$ with R value only 0.940, and the exclusion of every of the three other terms practically destroys the correlation as the R value diminishes to 0.88-0.92.

The analysis of the Eq. (3) shows that the growth of the solvent molecule dimensions reduces it penetration ability into polymers structure at the same time solvent nonspecific interaction with polybutadiene stimulated by its polarity and polarizability favours the penetration. The negative influence of solvent ability to the electrophilic solvation E_T is probably connected with the parallel appeared solvophobic effect. In addition we may point to that the E_T values for the all studied solvents excepting these for $CHCl_3$ and dioxane change in very narrow limits, namely 31.0-34.0.

The comparison of Eq. 2 and 3 which describe the regularities of polybutadiene swelling with the analogous equations for swelling of polyethylene which has a similar structure [4] shows on the proximity of behaviour of the both polymers – the signs and the validities of the separate terms of corresponding equations are the same.

For comparison in the Table 1 we give both the experimental values of lg S_m and these calculated by Eq. 3 as well as the divergencies of these values $\Delta lg\ S_M = lgS_{exp} - lgS_{calc}$. These deviations as a rule lie in the corridor of errors s±0.04 or only negligible leaves these limits.

One must mark that like the paper [4] the using of volume S_v or weight S_g parts of absorbed solvent gives the worses calculation results. For the all 13 solvent the corresponding R values are equal 0.885 or 0.840 only, and for 12 solvents after the exclusion of data for nitrobenzene correspondingly 0.995 or 0.968.

For the corroboration of the received conclusion, we made here analogical treatment of the data tooken from the paper [1] on the equilibrical swelling of samples of butylrubber (an copolymere of isobutylene with the little amount of isoprene) [1] vulcanized in the presence of ZnO, S and tetramethyltiuramdisulfide. The density of samples after the exhaustive extraction with acetone was equal 0.91 g/ml. The equilibrium of swelling process comes after 72 hours, the results presented in volume parts of absorbed solvent are recalculated like as in polybutadiene case in the mole/g S_m.

Also in this case the correlation primary reachable at the generalization of the data by means of Eq.1 is unsatisfactory (R=0.876) but for achievement of an acceptable value R ≥ 0.95 one must exclude from consideration the data for already two solvents from 18 namely $C_2H_4Br_2$ and $CHBr_3$. Thus we receive the follow equation:

$$lgS=-0.911 + (6.45 \pm 1.50)\ f(n^2) + (4.32 \pm 0.74)\ f(\varepsilon) - (3.47 \pm 0.25)\ 10^{-3}B$$
$$- (0.07 \pm 0.03)\ E_\tau - (1.69 \pm 0.93)10^{-3}\ \delta^2 - (3.62 \pm 0.54)10^{-3}\ V_M \qquad (4)$$

with R = 0.977, s ± 0.118 and r_1=0.176; r_2=0.549; r_3=0.804; r_4=0.664; r_5=0.302 and r_6=0.08.

Like to polybutadiene and polyethylene cases [4] the disposition of solvent to the specific solvation supresses their penetration into polymer hydrocarbonic structure which is obviously connected with its hydrophobicity and neutrality. At the same time the electrophilic solvent are as a rule hydrophilic. A negative influence on the solvent penetration have the factor of its molecule dimensions V_M too. The growth of solvent molecule polarity and polarizability which determines their ability to nonspecific solvation favours the swelling. The all above – named facts permit to suggest that the solvation process realized at swelling of polyethylene, polyisobutylene, and butylrubber are of the same kind.

The exclusion the cohesion energy term which is of little significance reduces the R-value only weakly to 0.972 and this way the dependence of butylrubber swelling degree from the solvent properties may be satisfactory described by the 5-parameter equation (5):

$$lgS_M=-0.07+(4.84\pm1.34)f(n^2)+(4.59\pm0.80)f(\varepsilon)-(3.55\pm0.27)10^{-3}B$$
$$-(0.10\pm0.02)E_T-(3.48 \pm 0.59)10^{-3}\ \delta^2 - (3.62 \pm 0.54)10^{-3}\ V_M \qquad (5)$$

with R = 0.972, s ± 0.10.

However the further exclusion of some equation terms lowers the correlation already essentialy or destroys it completely: R of the four – parameter equations without δ^2 and $f(n^2)$ is equal 0.947, without δ^2 and E_T 0.934, without δ^2 and V_M or $f(\varepsilon)$ 0.90 only.

In that way is confirmed the above-named opinion of the complexity and multilaterality of the processes realized in the polymer swelling. These may be quantitatively characterized only on condition of their combined examination which is reachable by means of LFER-equations.

In the Table 2 are given the values lg S_M of butylrubber calculated by the use of Eq.(5) and their deviations from experimental values Δlg S too.

Table 2. Experimental (According to [3]) and Calculated by Mmeans of Eq. 5 Amounts of Absorbed Solvents at the Swelling of Samples of Butylrubber

N	solvation	S_M	lgS_{exp}	lgS_{calc}	ΔS_{calc}
1	n-Pentane	0.0169	-1.7737	-1.6079	0.1658
2	n- Hexane	0.0231	-1.6398	-1.6361	0.0037
3	n-Heptane	0.0248	-1.6078	-1.6791	-0.0712
4	n-Decane	0.0136	-1.8665	-1.7927	0.0738
5	n-Hexadecane	0.0093	-2.0356	-2.0792	0.0435
6	Isooctane	0.0143	-1.8444	-1.6589	0.1855
7	Cyclohexane	0.0458	-1.3396	-1.4752	-0.1356
8	Methylcyclohexane	0.0365	-1.4517	-1.5251	-0.7334
9	Benzene	0.0194	-1.7144	1.7458	-0.0314
10	Toluene	0.0269	-1.5746	-1.7611	-0.1865
11	Dichloromethane	0.0446	-1.3503	-1.2990	0.5127
12	$CHCl_3$	0.0446	-1.3549	-1.4651	-0.1102
13	$CHCl_4$	0.0472	-1.3256	-1.4646	-0.1390
14	Bromomethane	0.0359	-1.4325	-1.2302	0.2024
15	1,2 –Dibromoethane		-2.2875	-2.7206	-0.1225
16	$CHBr_3$	0.0059	-2.2320	-2.4916	0.2310
17	Pyridyne	0.0025	-2.5981	-1.6386	0.6892
18	1,4 – Dioxane	0.0006	-2.7226	-1.4765	0.7556

REFERENCE

[1] Bristow I. M., Watson W.F.(1958)., *Trans. Faraday Soc.*, 54(12), P. 1731.

[2] Flory P.J. *J. Chem. Phys.* (1956), 18, P.108

[3] Aminabhavi T.M., Harogoppad S.B., Khinnavar R.S., Balundgi R.H.(1991), *JMS-Rev. Macromol.Chem. Phys.*, C 31(4), P.433

[4] R. Makitra, Ya. Pyrih, E. Zagladko, A. Turovskiy and G. Zaikov. *Journal of Applied Polymer Science* (2000).

[5] Scott R.L., Magat M. *J. Polymer Sci.* (1949), 4, P. 555.

[6] Quant. Struct. – *Act. Relat.* (1985), 4, P.29

Chapter 9

INDUCED DECOMPOSITION OF BENZOYL PEROXIDE BY 4-[4-(DIMETHYLAMINO)STYRIL]PYRIDINE AND ITS N-OXIDE

***M. A. Turovskyj,*[a] *I. O. Opeida,*[a] *O. V. Kusch,*[a] *G. D. Titskii*[a], *M. K. Turovska*[a] *and G. E. Zaikov*[b]**

[a]Institute of Physico-Organic and Coal Chemistry of the National Academy of Sciences of Ukraine,R. Luxemburg,70, 83114 Donetsk, Ukraine, e-mail:turov@dongu.donetsk.ua

[b]Institute of Biochemical Physics, Russian Academy of Sciences, 4ul. Kosygina, 117334 Moscow, Russian Federation, e-mail: chembio@sky.chph.ras.ru

ABSTRACT

Reactions of benzoyl peroxide with 4-[4-(dimethylamino)styryl]pyridine and its N-oxide have been studied in chlorobenzene within temperature range of 283-308 K. Under peroxide excess pseudo-first-order rate constants have been found, which are nonlinearly related on the peroxide concentration. This fact is explained by the formation of the complex of benzoyl peroxide and amines. Rate constants of decomposition of complexes and equilibrium constants of the complex formation have been determined. The presence of free radicals has been established in the reactions. The radical formation efficiency for reactions under investigation is about 60 %. The dominant effect of dimethylamino group in the acceleration of benzoyl peroxide decomposition has been shown.

Keywords: kinetics; 4-[4-(dimethylamino)styryl]pyridine and its N-oxide; benzoyl peroxide; polymerization of methyl methacrylate; initiation; chemiluminescence.

1 INTRODUCTION

The reactions of induced peroxide decomposition by closed-shell molecules is well known in peroxide chemistry [1-7]. Nucleophiles can react with peroxide substrates either by an single-electron transfer (SET) reaction or by nucleophilic displacement (S_N2) on the O-O bond of the peroxide. Distinguishing between S_N2 and SET paths is subtle and can be very difficult. As W.A. Pryor and W.H. Hendricson have noted [3], kinetics and the products of these reactions can be rationalized by both mechanisms, and more subtle factors such as solvent polarity effects, Hammett ρ values also do not allow distinction to be made. Even the accelerated production of free radicals cannot be taken as a conclusive proof for an SET mechanism, because S_N2-produced intermediate homolysis like the peroxide itself. The reaction of benzoyl peroxide with amines is a typical case. For example, the accelerated decomposition of benzoyl peroxide by N,N-dimethylaniline, which is model of biological oxidation, has been characterized as both an SET reaction[8] and an S_N2 reaction[2,9].

Such two-component systems as tertiary amines - peroxide use for the initiation of free radical reactions[10]. Our interest in the reactions of 4-[4-(dimethylamino)styryl]pyridine (DASP) and its N-oxide (DASPO) with benzoyl peroxide (BPO) originated with our observation that this systems gave a substantial yield of radicals and accelerated the polymerization of methyl methacrilate at ambient temperatures. We selected amines DASP and DASPO because the efficiencies of radical production for the reaction DASP and DASPO with BPO are 60-70 %. By comparison, the decomposition of BPO accelerated by N,N-dimethylaniline produces free radicals with efficiency of 10-20 % [9,10]. We also hoped this reactions would shed additional light on the problem of the mechanism of amine-induced reaction of peroxide.

2 EXPERIMENTAL

2.1 Materials

DASP and DASPO were prepared according to the literature procedure[11]. BPO was recrystallized twice from ethanol. Iodometric titration indicated that the peroxide was 99.4% pure. Chlorobenzene was purified according to standard technique and distilled prior use. Methyl methacrylate (MMA) was freed from inhibitor by washing with 5 % aqueous sodium hydroxide, dried over calcium chloride for 24 h and distilled under reduced pressure. 2,2,6,6-tetramethyl-1-piperidinyloxy was purified by sublimation. 1,10 – phenanthrolin complex of thionyltrifluoracetonate of europium was prepared as reported [12].

2.2 Kinetic Measurements

Pseudo-first-order rate constants (k_{obsd}) were determined by amines absorbance decrease at λ_{max}=375 nm for DASP (ε= 30000±400 $M^{-1}cm^{-1}$) and the consumption of DASPO at λ_{max}=395 nm (ε= 32000±500 $M^{-1}cm^{-1}$). Electronic spectra were recorded on a spectrophotometer "Specord UV VIS ". A dried and freshly distilled chlorobenzene was used as a solvent. The initial amine concentration was of 5×10^{-5} M. All reactions were carried out under peroxide excess (0.033-0.363 M) over the amine.

In a typical experiment the quartz 1-cm cells were filled with 2.8 mL of a solution of BPO and were placed in the thermostated compartment (283.0-308.0 ± 0.1 K) of the spectrophotometer. Then 0.2 mL of a solution of 5×10^{-5} M of DASP or DASPO was added and mixed well. Disappearance of the amines was monitored as a function of time. The cell of matching was filled with the solution of relevant BPO concentration.

The measurement of the rate of initiation of polymer chain in initiated polymerization of MMA by DASP - BPO and DASPO - BPO at 303 K in chlorobezene involves the use of the inhibitor 2,2,6,6-tetramethyl-1-piperidinyloxy. In the presence of this substance, there is an induction period in the polymerization during which the inhibitor is consumed. Reactions of polymerization of methyl methacrylate were carried out up to a monomer conversion of 3%. The peroxide-amine induced polymerization of MMA has been studied kinetically by the dilatometric method. The dilatometric measurements were made with sufficient frequency.

Intensity of a chemiluminescence was measured by the photoelectric multiplier FEU-38 with an electrometric signal amplification. We used 1,10 - phenanthrolin complex of thionyltrifluoracetonate of europium as an activator. The reaction was carried out under pseudo-monomolecular conditions at 303 K under peroxide excess. Assuming that the luminescence intensity change reflects the initiator concentration change adequately, a rate constant of the initiator decomposition (k_{chem}, s^{-1}) was calculated from ln (I/I_0) = f(t) dependence.

3 Results and Discussion

The consumption of the amines DASP and DASPO in reactions with benzoyl peroxide obeys the kinetic equation of the first order. The experimental data are shown in Fig. 1.

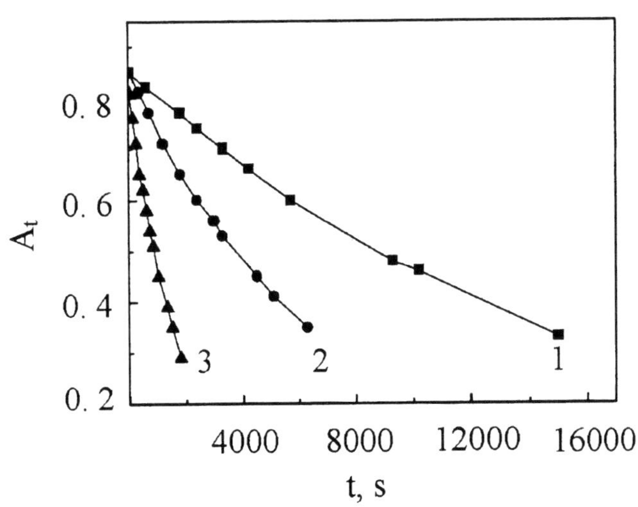

Figure 1. Dependence of the absorbance of DASP on time at 298 K for the reaction of DASP with BPO. $[DASP]_0 = 5\times10^{-5}$ M; [BPO] = 2.6×10^{-2} (1); 6.5×10^{-2} (2); 2.5×10^{-1} M(3)

The change of the absorbance with time, generally, is represented by the equation:

$$A_t = A_\infty + (A_0 - A_\infty)\exp(-k_{obsd} \cdot t) \quad (1)$$

where A_0 is the absorbance at the beginning of reaction, A_∞ and A_t are absorbances at "infinity" and variable times, respectively.

The statistical analysis of the experimental data has shown, that for A_∞ parameter values the p-level defining statistical significance are within an interval of 8×10^{-4}-8×10^{-1}, therefore in further calculations we used equations without this parameter. The direct experimental estimation of A_∞ also showed that it was< 0.04 under conditions of our experiment.

The satisfactory accordance between the experimental and calculated values of the amine absorbance is observed at different concentrations of peroxide within investigated temperatures (Fig.1). The points on the plots are experimental values of the absorbance; the solid lines are calculated values of the absorbance of DASP according to the equation 1.

The analysis of the kinetic data given below has shown, that the observed rate constants of the reaction of DASP with BPO do not depend on the initial amine concentration (Table 1).

Table 1. Pseudo-First-Order Rate Constants (k_{obsd}) for Reaction of BPO with 1 at Different Concentrations of DASP at 298 K, ([BPO] =0.26 M)

$[DASP]_0\times10^5$ (M)	$k_{obsd.}\times 10^4$ (s^{-1})
1.46	6.3±0.2
2.7	5.5±0.4
3.27	5.6±0.1
4.38	5.6±0.2

Dependencies of $k_{obsd.}$ on the concentration BPO are nonlinear at all temperatures (typical dependence is given in Fig. 2.)

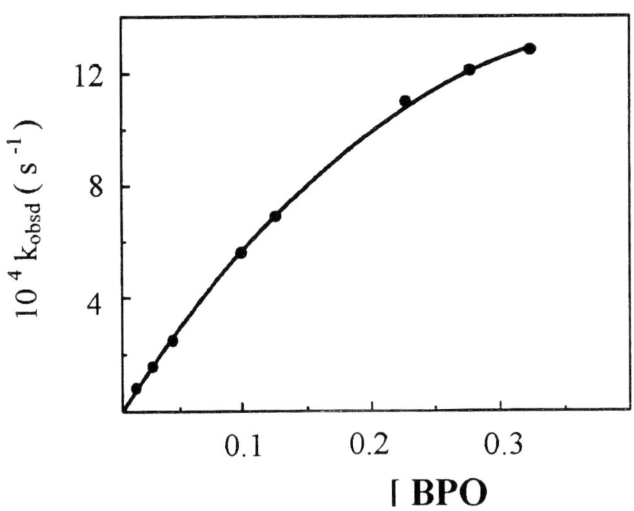

Figure 2. Dependence of k_{obsd} on benzoyl peroxide concentration for the reaction of DASP with BPO at 308 K and $[DASP]_0=5\times10^{-5}$ M

The deviation from linearity of the observed dependencies can be explained by the formation of the BPO - amine complex. In this case, the investigated reactions could be described by the Scheme 1:

Scheme 1

In Sheme 1: X - $CH=CH-C_5H_4N$ (DASP) and $CH=CH-C_5H_4N\rightarrow O$ (DASPO); K_C (M^{-1}) is the equilibrium constant of the BPO complex formation with DASP or DASPO, k_d (s^{-1}) is the rate constant of decomposition of the complex [BPO··· DASP] or [BPO··· DASPO].

According Sheme 1, the observed rate constant k_{obsd} can be calculated

$$k_{obsd} = k_d K_C[BPO] / (1+K_C[BPO]) \quad (2)$$

In the case of the equilibrium is estableshed quickly k_d and K_C may be calculated using the following equation:

$$A_t = A_0 \times \exp(-k_d K_C[BPO]\, t /(1+K_C[BPO])) \quad (3)$$

k_d and K_C can be presented as:

$$k_d = \exp\ (\ln A - E/RT) \quad (4)$$

$$K_C = \exp\ (\Delta S/R - \Delta H/RT) \quad (5)$$

where A is the pre-exponential factor; R is the gas constant; E is the activation energy of a complex decomposition; T - thermodynamic temperature; ΔH is the enthalpy of a complex formation; ΔS - the entropy of a complex formation.

So, using the equations 1, 4, 5 one can obtain

$$A_t = A_0 \times \exp(-\exp(\ln A - E/RT)\exp(\Delta S/R - \Delta H/RT)\ [BPO]t /(1 + (\Delta S/R - \Delta H/RT)^{\cdot}\ [BPO])) \quad (6)$$

This equation was used to handle simultaneously all available experimental data obtained in temperature range 283-308 K and at different peroxide concentrations, 241 experimental points of 6 runs. The calculated activation parameters are given in Table 2. Values of the p-level for all parameters of the equation 6 are less than 0.02, correlation coefficient is 0.995 (Fig. 3).

Table 2. Activation Parameters of Reactions of BPO with DASP and DASPO

Am	T(K)	$k_d \times 10^3$ (s^{-1})	K_C (M^{-1})	ΔH (kJ mol^{-1})	ΔS (J mol^{-1} K^{-1})	E_a (kJ mol^{-1})	Log A (A, s^{-1})
	288	0.96±0.09	0.96±0.09				
DASP	298	3.1±0.8	0.77±0.19	-16.1±0.2	-56.3±0.8	83.1±0.4	12.0±0.2
	308	9.1±3.3	0.62±0.19				
	283	1.08±0.07	1.02±0.07				
DASPO	298	4.6±1.6	0.83±0.3	-10.4±0.4	−36.6 ± 0.1	65±2	9.0±0.4
	308	9.9±2.2	0.71±0.17				

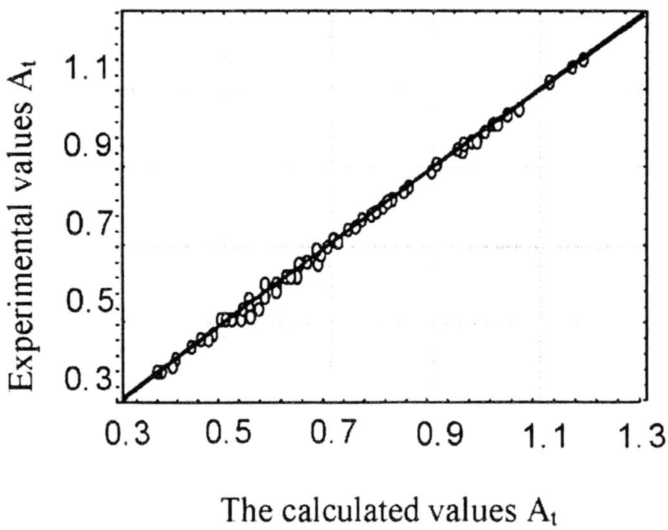

Figure 3. Comparison of experimental values of the absorbance of DASP in the reaction with BPO and calculated ones by equation 6

It worth to notice that the correctness of obtained parameters may be confirmed by comparison of values of $k_d \times K_C$, calculated with the data of Table 2 and the tangent of the curve of k_{obsd} on [BPO] at [BPO] $\rightarrow$ 0. For example, the value of $k_d \times K_C$ for the reaction BPO with DASP at 308 K equals to (5.9±1.3)×10^{-3} $M^{-1}s^{-1}$ (see Table 2) and the tangent of the curve of k_{obsd} on [BPO] (Fig. 2) equals to (5.50±0.05)×10^{-3} $M^{-1}s^{-1}$.

The decomposition of complexes DASP and DASPO with BPO occurred with significantly smaller activation energy, than BPO thermolysis. The rate constant of BPO thermolysis is $k = 4.9 \cdot 10^{14} \exp(-13100/RT)$.

Aminostyrylpyridines DASP and DASPO are polyfunctional compounds. Their molecules include pyridine-, N-oxide pyridine-, dimethylamino- and styryl- structural fragments. Each of them can potentially promote the decomposition of peroxide. To identify the role of N-oxide and styryl groups, kinetics of N-oxide styrylpyridine (SPO) reaction with BPO was studied ([BPO]=0.26 M, $[SPO]_0 = 4 \times 10^{-4}$ M, λ_{max} - 357 nm, ε = 22500 $M^{-1} cm^{-1}$) in chlorobenzene at 298 K. It was established, that the observed rate constant of this reaction

(k_{obsd} = 3,4×10^{-5} s^{-1}) is lower than the rate constant of a compound DASPO with BPO (k_{obsd} = 8×10^{-4} s^{-1}) under the same conditions. The higher reactivity of DASPO indicates the dominant effect of the dimethylamino group in the activation of benzoyl peroxide decomposition.

This fact allows to propose that amines DASP and DASPO, which are derivatives of N,N-dimethylaniline, react with BPO according [3]

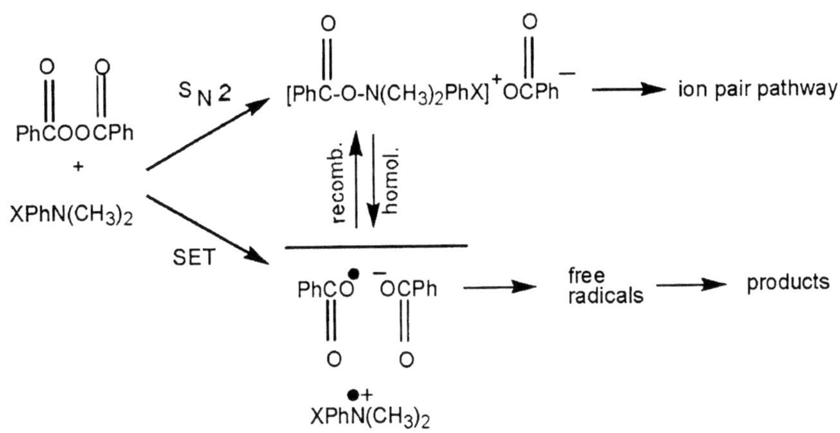

Scheme 2

We have detected the benzoic acid (1 mol per 1 mol of decomposed peroxide) in reaction products by the gas chromatography.

Scheme 2 illustrates the difficulty of experimentally distinguishing between S_N2 and ET mechanisms. We have not favored an S_N2 or SET mechanisms for this system, because we can attribute radical formation to the homolysis of salt.

The formation of radicals in BPO reaction with DASP was confirmed by monitoring a chemiluminescence as a result of the recombination of $RO_2^\bullet$ radicals in the reaction of homolytic oxidation of cumene initiated by BPO- DASP system. We observed fast increase of the chemiluminescence intensity at the moment of adding of **1** in the reaction vessel containing solution of the peroxide in cumene at 303 K. The intensity decreased with time exponentially. The value of BPO decomposition rate constant obtained by method of the chemiluminescence in the presence of DASP (k_{obsd} = (9.7 ±0,3)×10^{-4} s^{-1} at [BPO)] = 0.29 M) satisfactorily agreed with the value of k_{obsd} = (8,7± 0,4) ×10^{-4} s^{-1}, measured by UV-spectroscopy.

Methyl methacrylate polymerization in the presence of the inhibitor has been used to determine radical yields in the DASP -BPO and DASPO -BPO systems.

It has been established that the period of induction of MMA polymerization depends on concentration of the inhibitor linearly. The rate of polymerization after complete consumption of the inhibitor is taken to be equal to the rate of polymerization with inhibitor absent. It has been assumed that one molecule of inhibitor stops one radical chain [13] (Fig.4).

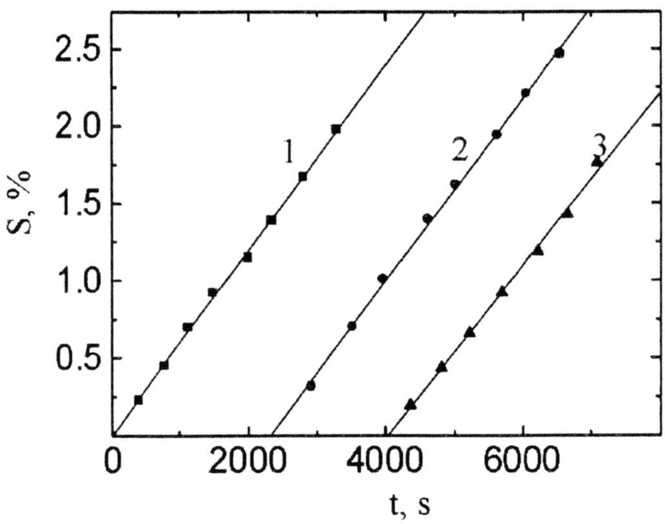

Figure 4. Polymerizations of MMA in chlorobenzene at 303 K in the presence of system DASPO -BPO and inhibited by 2,2,6,6-tetramethyl-1-piperidinyloxy. 1- polymerization with inhibitor absent; 2 - [In] = $3\cdot10^{-5}$; 3 - $5\cdot10^{-5}$ M. $[DASPO]_0 = 2.2\times10^{-4}$ M, $[BPO] = 7.5\times10^{-3}$ M, [MMA] = 25 vol. %

In eq. 6 f is the efficiency of radical production in the reactions of amine with peroxide.

$$W_{in} = 2fW_d = [In]/\tau \quad (6)$$

By combining this equations we find that

$$f = W_{in}/2W_d, \quad (7)$$

where W_{in} - initiation rate, τ - induction period, [In] - initial concentration of inhibitor, W_d - the rate of induced peroxide decomposition by DASP and DASPO. The initial concentrations of reagents were as follows: $[DASP]_0 = 2.5\times10^{-4}$ M, $[DASPO]_0 = 2.2\times10^{-4}$ M, $[BPO] = 7.5\times10^{-3}$ M, $[In]_0 = (0\text{-}5)\times10^{-5}$ M, [MMA] = 25 vol. %.

Values of f calculated from eq. 7 for systems DASP -BPO and DASPO -BPO are collected in Table 3.

In the paper[14] it was shown, that ionization potentials (*IP*) of donor molecules made considerable contribution to their reactivity in their reactions with peroxides. There are no still data on *IP* for compounds under investigation, therefore we have calculated the electron structure of DASP, DASPO, SPO *trans*-isomers by AM1 method. The ionization potentials were determined according to the Koopmans theorem. It can be seen in Table 4 that the calculated values of *IP* for compounds DASP, DASPO, SPO increase with the decrease of their reactivity in reactions with benzoyl peroxide.

Table 3. Radical Production in Reactions of BPO with DASP and DASPO Measured by Methyl Methacrylate Polymerization at 303 K in Chlorobenzene

Am	$[In]\times10^5$ (M)	τ (s)	$W_{pol}\times10^5$ (Ms^{-1})	$W_{in}\times10^8$ (Ms^{-1})	$W_d\times10^9$ (Ms^{-1})	F
DASP	0	0	1.32±0.01	–	8.9±0.1	0,70±0.18
	1	750±170	1.32±0.04	1.32±0.30		
	3	2600±160	1.29±0.03	1.16±0.07		
DASPO	0	0	1.41±0.01	–	10.5±0.2	0,60±0.04
	3	2300±110	1.39±0.01	1.29±0.06		
	5	4100±170	1.37±0.01	1.22±0.05		

Table 4. Pseudo-First-Order Rate Constants (k_{obsd}) for Reactions of BPO with Amines at 298 K ([BPO] =0.26 M) and Ionization Potentials (IP) of Amines

Am	$k_{obsd}\times10^4$ (s^{-1})	IP (eV)
$Me_2NC_6H_4CH=CHC_5H_4N\rightarrow O$	8.3±0.1	8.04
$Me_2NC_6H_4$-$CH=CH$-C_5H_4N	5.6±0.2	8.13
C_6H_5-$CH=CH$-$C_5H_4N\rightarrow O$	0.34±0.01	8.51

4 Conclusion

Polymerization of MMA in the presence of the inhibitor and the chemiluminescence at the oxidation indicate that free radicals are produced in DASP -BPO and DASPO -BPO reactions with high yield. Kinetic analysis of induced decomposition of BPO by amines DASP and DASPO allows to establish that these reactions involve the equilibrium stage of complex formation. The rate of unimolecular decomposition of amine – peroxide complex is about a 3-fold higher of the rate of BPO-thermolysis. Higher electron-donating properties of DASPO determine its high reactivity.

References

[1] Yablokova, N.V., Alexandrov, Y. A., *Uspechi Chimii.* **58** (1989) 908.
[2] Srinivas, S., Taylor, K.G., *J. Org. Chem.*, **55** (1990) 1779.
[3] Pryor, W. A., Hendricson, W.H., *J. Amer. Chem. Soc.*, **105** (1983) 7114.
[4] Pryor, W. A., Hendricson, W.H., *Tetrahedron Letter.*, **24** (1983) 1459.
[5] Gragerov, I. P., Skrunz, L.K., Geller, B. A., *Uspechi Chimii.*, **51** (1982) 119.
[6] Baj, S., Chrobok, A., *Polish. J. Chem.*, **72** (1998) 1131.
[7] Vidal, S., Court, J., Bonnier, J.M., *Tetrahedron.*, **40** (1984) 4911.
[8] Filliatre, C., Lalande, R., Pometan, J-P., *Bull. Soc. Chim. Fr.*, (1975) 1147.
[9] Walling, C., Indictor, N. J., *J. Amer. Chem. Soc.*, **80** (1958) 5814.
[10] O Driscoll, K. F., Lyons, P.F., Patsiga, R. J. *Polym. Sci.*, **3** Part A. (1965)1567.

[11] Parker, E.D. Furst, A., *J. Org. Chem.,* **23** (1958) 201.
[12] Kononenko, L.I., Tischenko, M. A., Vitkun, R.A., Poluektov, N. S., *J. Inorg. Chim.*, **10** (1965) 2465.
[13] Plusnin, A.N., Chirkov, N.M., *Teor. eksp. khim.*, **2** (1966) 777.
[14] Beylerian, N.M., *Kinet. Katal.*, **38** (1997) 32.

INDEX

3

A

B

C

D

E

F

G

H

I

J

K

L

M

N

O

P

Q

R

S

T

U